SELECTED SCREENING TESTS

FOR GENETIC METABOLIC DISEASES

SELECTED SCREENING TESTS
FOR
GENETIC
METABOLIC DISEASES

GEORGE H. THOMAS, Ph.D.

Assistant Professor of Pediatrics and Medicine, The Johns Hopkins University School of Medicine
Director, Genetic Laboratory, The John F. Kennedy Institute, Baltimore, Maryland

AND

R. RODNEY HOWELL, M.D.

Professor and Chairman, Department of Pediatrics,
The University of Texas Medical School at Houston, Houston, Texas

YEAR BOOK MEDICAL PUBLISHERS • INC.

35 EAST WACKER DRIVE • CHICAGO

Library of Congress Catalog Card Number: 72-95730

International Standard Book Number: 0-8151-4704-4

Preface

This book represents the collection of screening methods for the detection of certain genetic metabolic diseases that have been widely used at The Johns Hopkins Hospital for a number of years. These methods were usually set up in response to a specific need, i.e., for the study of patients who were thought clinically to be good candidates for genetic metabolic disease. They were collected and organized largely since the opening of The John F. Kennedy Institute when an expanded program for the detection of metabolic diseases was introduced.*

Most of the methods are not original and represent the efforts of many distinguished workers in this field over many years. We have made modifications in some of these standard methods and in many instances have included data derived from our own efforts.

It is hoped that in the not too distant future, one might be able to analyze a wide variety of biological fluids for all known metabolites by a single, all-inclusive technique. This would be optimal and would remove the necessity for the rather "hit or miss" methods that one is currently forced to utilize.

Unfortunately, at the present time, we are still restricted in our vision by the chemical nature of our various screening methods. For this reason we feel strongly that for a laboratory screening program to be most effective it must be in the "main stream" of clinical practice, being closely connected with professional persons having expertise in both the clinical and biochemical manifestations of genetic metabolic disease. It is only in this manner that both the laboratory staff and the clinician can best contribute to the detection and possible treatment of individuals with genetic metabolic disease.

We would like to acknowledge the technical assistance of Mr. Joseph Bace for the performance of many of the laboratory methods upon which this manual is based. In addition, certain of the tests were worked out by Mrs. Johanna Harris to whom we are also indebted. We have benefited from the experience of Dr. Roger Stevenson in the utilization of the single dimensional method for amino acid chromatography on paper and from the opportunity to work with Dr. Neil Holtzman in the evaluation of biological fluids for the detection of metabolic disorders.

In addition, we would like to express our appreciation to Dr. Victor McKusick who has made available a large number of patients with unusual metabolic problems. Finally we would like to acknowledge the fact that the hospital house staff has, over the years, challenged our imagination by obtaining urine specimens for examination containing not only a wide variety of colors and crystals but also an almost endless variety of unusual odors. Their interest in this area of human disease has certainly served to accelerate the growth of our screening program.

*Supported by Project Grant #917 from Maternal and Child Health Services.

We are also indebted to Mrs. Kathryn W. Johnson and Mrs. Carol Miller, who have been particularly helpful in proofreading, and to Mrs. Beverly Cogdill for the manuscript preparation.

We are also aware of a growing number of clinical investigators throughout the world who are performing similar screening tests and who, in the process, are obtaining new data. We would appreciate hearing comments from these investigators regarding both the methods that are included in this book, as well as any additional ones that have proved to be of value for the detection of genetic metabolic disease. Through this mechanism it is hoped that we can draw on the experience of others in keeping this manual updated and useful.

George H. Thomas, Ph.D. and R. Rodney Howell, M.D.
Baltimore, Maryland and Houston, Texas

Table of Contents

CHAPTER 1

Screening Tests: General Discussion

In the 1971 edition of his catalogue of genetically determined phenotypes in man, McKusick lists some 1,876 variations in man, for which there is evidence of a major genetic influence (1). Moreover, he also estimates that the number of genetic loci, identified by this catalogue of disorders, is only a very small portion of the total number of gene loci in the total human chromosome complement—perhaps about 1%.

The Problem

Certainly not all of these known genetic variations are associated with disease; however many, if not most, are associated with severe lifelong handicaps; frequently, mental retardation. Clow et al. indicate that about 6% of the admissions to pediatric hospitals are due to disorders which are clearly genetic, while another 15% of the admissions are for conditions which are thought to have at least some genetic component (2). As the deaths resulting from causes such as the viral exanthems are reduced by advances in medical care, the percentage of deaths resulting from causes which are solely or partially genetic in origin increases (3,4).

From the standpoint of medical practice these figures serve to emphasize the magnitude of the role that genetic variation plays as a causative factor in human disease, both in terms of well-documented disorders as well as, most likely, an even larger number of, as yet, unrecognized disorders. The recognition that the medical burden, due to genetically related disorders, is both considerable and increasing (as a percentage of the total although not in absolute numbers) combined with increased knowledge regarding the prevention and treatment of genetic disorders has placed increased emphasis on the diagnosing of affected individuals.

Individuals with genetic disease can often be difficult to detect. Most of the genetic disorders are individually rare; frequently they are difficult to diagnose by clinical examination alone. The development of a variety of screening programs for the detection of certain genetic disorders has thus been necessary.

Philosophical Aspects of Screening

Screening for genetic-metabolic disease demands certain philosophical considerations, centering around the investment in the activity undertaken and the potential gain to the patient and/or his family for whom the test is being performed. For a more detailed consideration of this area, the reader is referred to a discussion by Scriver (5). As summarized in the World Health Organization Technical Report Series No. 401 (6), screening programs for genetic disorders may be regarded in the following manner:

1

1. ". . . . primarily as an activity in preventive medicine aimed at detecting individuals in whom it is likely that a specific hereditary disease will develop for which preventive or therapeutic measures are available."

2. ". . . . as contributing to the collection of scientific genetic information on a population as well as to immediate public health practice."

It might be added that, with the advent of the technique of intrauterine diagnosis, the prevention aspect, in many cases, may not be the prevention of the disease process in the index case, but rather the prevention of the same disease in, as yet, unborn sibs by techniques which ". . . allow carriers for a given variant gene to make informed choices regarding reproduction. . ." (7). It is still an unfortunate fact that, with a few exceptions, the techniques of precise genetic counseling and/or amniocentesis can only be offered to those families in which a high risk for a specific genetic disease has been established by the diagnosis of a genetic disorder in an affected member.

Medical Aspects of Screening

Various considerations which are important in determining the scope of any screening program have been reviewed in the World Health Organization report referred to earlier (6). Some of the factors listed in this report which will or should influence the design of any screening program are "The availability of facilities for the validation of the presumptive diagnosis afforded by the screening test; the amenability of the disorder to treatment; and the economic and social effects of diagnosing such disorders, both on the family and on the community."

Various aspects of the above considerations have been reviewed in some depth by a number of individuals. Special areas which have been discussed in detail include the medical management of detected patients (2), ethical issues (7,8), certain social and genetic aspects (9), dietary treatment of metabolic defects (10) and intrauterine diagnosis (11,12).

Laboratory Aspects of Screening

In addition to the above considerations, in deciding to introduce a screening program one must determine in detail how he will actually carry out the detection process. Various programs introduced in the past have utilized a number of approaches including some of the following:

1. Screening for the detection of persons heterozygous for a particular gene, e.g., carriers of a gene causing Tay-Sachs disease or a gene for sickle disease.

2. Screening of all newborns for the detection of infants homozygous for genes causing certain disorders that are most often treatable, e.g., persons homozygous for genes causing phenylketonuria or genes for the galactosemias.

3. Screening of individuals in certain "high risk groups," e.g., individuals having one or more of the following: mental retardation, delayed milestones, seizures, failure to thrive, ammonia intoxication, unexplained ketoacidosis, evidence of central nervous system dysfunction or a relative with a specific hereditary metabolic disorder.

The screening tests described in this manual are primarily concerned with the detection of a variety of inborn errors of metabolism in "high risk patients" of the type listed in group 3 of the previous paragraph. In general, screening programs of this type are based on several simple chemical methods capable of detecting alterations in the biochemical composition of extracellular fluid—usually plasma and urine (13-17). Several such programs have been described in some detail and, along with our own experience, have served as the source of information for this manual (13, 14, 18-31).

Regardless of the tests employed, it is essential that one be aware of the fact that a positive screening test result does not pinpoint the enzymatic defect but rather serves to indicate one or more general metabolic pathways in which a biochemical alteration may be located. In addition, while many of the tests outlined in this manual are extremely simple, have withstood the test of time and are inexpensive to perform, they still require considerable skill and experience on the part of the laboratory technician to ensure a consistently correct interpretation of the results. Important decisions are frequently made on the basis of these screening tests and it is imperative that the person performing the tests know what constitutes a negative result, what constitutes a positive result, and be fully aware of the various problems that can lead to erroneous interpretations on both sides of the scale.

"Follow-up" Requirements of Screening

For the above reasons, our experience indicates that laboratory screening programs should be closely integrated into the overall clinical program in such a way as to encourage a maximum exchange of information between the clinician and the laboratory staff. A close working relationship of this type helps reduce the possibility of a patient being carried with a vague diagnosis of some metabolic disorder based on a "positive screening" test result without proper additional confirmation tests.

In this regard, it is essential to emphasize, once again, that virtually all of the tests outlined in this book are screening tests. For example, no diagnosis should be made on the basis of a positive ferric chloride test regardless of how "typical" it may appear. Final diagnosis should be made only on the basis of highly sophisticated and quantitative methods, such as quantitative amino acid determinations and/or tissue enzyme analyses carried out under optimal conditions. On occasion, there has been difficulty in obtaining accurate back-up laboratory tests, but over recent years considerable efforts have been made toward regionalization and nationalization of certain genetic laboratory facilities. The National Genetics Foundation, Inc., 250 West 57th Street, New York City, New York, will provide, for example, referral service for physicians for any sophisticated laboratory tests needed to establish a genetic diagnosis. The National Foundation-March of Dimes (1275 Mamaroneck Ave., White Plains, New York 10605) publishes an extensive genetic services directory.

References

1. McKusick, V. A.: *Mendelian Inheritance in Man, Catalogs of Autosomal Dominant, Autosomal Recessive and X-Linked Phenotypes* (3d ed.; Baltimore: Johns Hopkins Press, 1971).

2. Clow, C. L., Reade, T. M., and Scriver, C. R.: Management of hereditary metabolic disease. The role of allied health personnel, New England J. Med. 284:1292-1298, 1971.

3. Roberts, D. F., Chavez, J., and Court, S. D. M.: The genetic component in child mortality, Arch. Dis. Childhood 45:33-38, 1970.

4. Carter, C. O.: Changing patterns in the causes of death at the Hospital for Sick Children, Great Ormond St. J. 11:65-68, 1956.

5. Scriver, C. R.: Screening newborns for hereditary metabolic disease, Pediat. Clin. North America 12:807-821, 1965.

6. World Health Organization Scientific Group: Screening for inborn errors of metabolism, *Technical Report Series* 401:1-57, 1968.

7. Institute of Society, Ethics and the Life Sciences: Ethical and social issues in screening for genetic disease, New England J. Med. 286:1129-1132, 1972.

8. Littlefield, J. W.: Genetic screening (editorial), New England J. Med. 286:1155-1156, 1972.

9. Stevenson, R. E., and Howell, R. R.: Some medical and social aspects of the treatment for genetic-metabolic diseases, Ann. Am. Acad. Political Social Science 399:30-37, 1972.

10. Holtzman, N. A.: Dietary treatment of inborn errors of metabolism, Ann. Rev. Med. 21:335-356, 1970.

11. Motulsky, A. G., Fraser, G. R., and Felsenstein, J.: Public health and Long Term Genetic Implications of Intrauterine Diagnosis and Selective Abortion in Bergsma, D. (ed.): *Intrauterine Diagnosis,* Birth Defects Original Article Series Vol. VII, no. 5 (New York: The National Foundation, 1971).

12. Milunsky, A., Littlefield, J. W., Kanfer, J. N., Kolodny, E. H., Shih, V. E., and Atkins, L.: Prenatal genetic diagnosis, New England J. Med. 283:1370-1381; 1441-1447; 1498-1504, 1970.

13. Hsia, D. Y-Y., and Inouye, T.: *Inborn Errors of Metabolism: II. Laboratory Methods* (Chicago: Year Book Medical Publishers, Inc., 1966).

14. Tocci, P. M.: The Biochemical Diagnosis of Metabolic Disorders by Urinalysis and Paper Chromatography in Nyhan, W. L. (ed.): *Amino Acid Metabolism and Genetic Variation* (New York: McGraw-Hill Book Company, Inc., 1967).

15. Howell, R. R.: Diagnostic Procedures for Genetic Metabolic Defects in Bartalos, M. (ed.): *Genetics in Medical Practice* (Philadelphia: Lea & Febiger, 1968).

16. Snyderman, S. E.: Diagnosis of metabolic disease, Pediat. Clin. North America 18:199-208, 1971.

17. Thomas, G. H., and Scott, C. I.: The laboratory diagnosis of genetic disorders, Pediat. Clin. North America (in press).

18. Berry, H. K.: Detection of metabolic disorders among mentally retarded children by means of paper spot tests, Am. J. Ment. Deficiency 66:555-560, 1962.

19. Efron, M. L., Young, D., Moser, H. W., et al.: A simple chromatographic screening test for the detection of disorders of amino acid metabolism. A technique using whole blood or urine collected on filter paper, New England J. Med. 270:1378-1383, 1964.

20. Scriver, C. R., Davies, E., and Cullen, A. M.: Application of a simple micromethod to the screening of plasma for a variety of aminoacidopathies, Lancet 2:230-232, 1964.

21. Renuart, A. W.: Screening for inborn errors of metabolism associated with mental deficiency or neurologic disorders or both, New England J. Med. 274:384-387, 1966.

22. Perry, T. L., Hansen, S., and MacDougall, L.: Urinary screening tests in the prevention of mental deficiency, Canad. M. A. J. 95:89-95, 1966.

23. Buist, N. R. M.: Set of simple side-room urine tests for detection of inborn errors of metabolism, Brit. M. J. 2:745-749, 1968.

24. Berry, H. K., Leonard, C., Peters, H., *et al.*: Detection of metabolic disorders—chromatographic procedures and interpretation of results, Clin. Chem. 14:1033-1065, 1968.

25. Verma, I. C.: Simple urine tests for screening for inborn errors of metabolism, Indian J. Pediat. 37:487-491, 1970.

26. O'Brien, D.: *Rare Inborn Errors of Metabolism in Children with Mental Retardation*, PHS Publication no. 2049 (Washington, D.C.: U.S. Government Printing Office, 1970).

27. Mabry, C. C.: High voltage electrophoresis, CRC Crit. Rev. Clin. Lab. Sci. 1:135-190, 1970.

28. Duncan, I. W.: Urinary screening tests to detect metabolic disorders, Alaska Med. Sept., 1970, pp. 86-89.

29. Scriver, C. R., Clow, C. L., and Lamm, P.: Plasma amino acids. Screening, quantitation, and interpretation, Am. J. Clin. Nutrition 24:876-890, 1971.

30. Saifer, A.: Rapid screening methods for the detection of inherited and acquired aminoacidopathies, Advances Clin. Chem. 14:145-218, 1971.

31. Boggs, D. E.: Detection of inborn errors of metabolism, CRC Crit. Rev. Clin. Lab. Sci. 2:529-572, 1971.

CHAPTER 2

Examination of Urine: Odors, Crystals and Colors

The presence of an abnormally large amount of one compound in the urine resulting from a metabolic defect often can be detected by using one's ordinary unaided senses. It is thus an important first step that one routinely examine urine samples for the presence of any unusual odors, color, or crystals.

Urinary Odors

The presence, for example, of a distinctive odor during the first days of life is often the first clue that an infant is the victim of an inborn error of metabolism (1-4). Those disorders which are now known to be associated with the presence of odorous substances include inborn errors of certain amino acids as well as certain fatty acids (Table 2-1).

The odor of phenylacetic acid, which is excreted in increased amounts in the urine of phenylketonuric patients, can be particularly valuable when the urine sample has not been well preserved, since the phenylpyruvic acid may have disappeared thus yielding a negative ferric chloride test. Likewise, in our experience, the odor of maple syrup urine disease is diagnostic of this condition and may well be the first clue as to the nature of the problem in a severely ill infant (5). We have noted that the presence of the odor of maple syrup urine disease and phenylketonuria is most prominent when the urine is smelled after the performance of the

TABLE 2-1.—Inherited Disorders Associated With a Specific Odor

Disorder	Compound	Odor	Reference
Phenylketonuria	Phenylacetic acid	Musty odor	(4)
Maple syrup urine disease	Branched chain a-keto acids	Maple syrup or burned sugar	(5)
Isovaleric acidemia	Isovaleric acid	Cheesy or sweaty feet	(6,7)
Oasthouse disease*	a-Hydroxybutyric acid	Oasthouse or brewery	(14)
Methionine malabsorption*	a-Hydroxybutyric acid	Oasthouse	(15)
Hypermethioninemia	a-Keto-γ-methiol butyric acid	Rancid butter or rotten cabbage	(16)
Butyric/hexanoic acidemia	Butyric and hexanoic acids	Sweaty feet	(8,9)
Trimethylaminuria	Trimethylamine	Stale fish	(17)

*It has been suggested the methionine malabsorption syndrome and the oasthouse disease may be identical disorders (18).

dinitrophenylhydrazine reaction. This probably results from the strong acidification and volatilization of the materials responsible for the odor.

The detection of the odor in the inborn errors of fatty acid metabolism, e.g., isovaleric acidemia (6,7) and butyric/hexanoic acidemia (8,9) is especially important as there are, at the present time, no simple screening tests for this group of disorders.

As odor is a very subjective quality, the description of smells is of necessity very imprecise. One, therefore, should not dismiss an odor as unimportant simply because it does not "fit" the list of well-recognized disorders associated with odors. Instead, the prolonged presence of any strong unusual odor in an infant should lead to a complete biochemical investigation in an attempt to identify the volatile compound.

Urinary Crystals

In addition to noting the odor, one should also examine the urine for the presence of crystals (Table 2-2). Although it is extremely common for urine samples to have a considerable amount of precipitate after refrigeration, the presence of large quantities of crystalline material should always arouse suspicions, and such conditions as cystinuria and orotic aciduria should be considered. Indeed the first patient with orotic aciduria was diagnosed due to the awareness of the laboratory personnel that there were enormous numbers of needle-like crystals present in the urine (which ultimately proved to be orotic acid) (10).

A clue as to the nature of the crystals might be obtained from the routine screening tests given in this manual. For example, the presence of abnormally large amounts of cystine will result in a positive nitroprusside test (Chapter 6), while an increased concentration of tyrosine will result in a positive result with the nitrosonaphthol test (Chapter 5).

If the usual screening tests are negative in urines containing large numbers of crystals, a number of additional inborn errors of metabolism associated with crystaluria should be considered. The crystals can be simply dissolved in an alkaline or an acid medium and spectral determinations can be run to see whether or not the compound is a purine or pyrimidine as a first approximation.

While it is true that uric acid crystals occur in normal urine, the presence of unusually large amounts of these crystals can be the first evidence for the presence of an inherited defect in the metabolism of uric acid. Buist, for example, cites a case of familial hyperuricemic neuropathy which was diagnosed in this fashion (3).

TABLE 2-2.—Inherited Disorders Which May Be Associated
With Urinary Crystals (3,4)

Disorder	Compound	Crystals
Orotic aciduria	Orotic acid	Fine spindles
Lesch-Nyhan syndrome	Uric acid	Various forms
Cystinuria	Cystine	Hexagonal
Xanthinuria	Xanthine	Lemon-shaped
Tyrosinemia	Tyrosine	Feathery
Hyperoxaluria	Calcium oxalate	Ditetragonal pyramid crystals
Fanconi's syndrome	Leucine	Circular

Urinary Colors

The problem of urines presenting various colors is a constant one (11). Persons with rare conditions such as McArdle's disease (muscle phosphorylase deficiency) will report reddish brown urine after extreme exercise. This color is caused by the presence of myoglobin in the urine. A variety of other conditions may also produce myoglobinuria and hemoglobinuria.

A familial disorder, thought to be associated with a defect in tryptophan absorption, has been found to be associated with a bluish coloration of diapers. This, the so-called "blue diaper syndrome," is thought to be caused by indican formed by bacterial action on the unabsorbed tryptophan (12). The complaint that urine darkens on standing, particularly when it is alkaline, or that there is dark staining of the underwear should suggest the presence of homogentisic acid in urine (as in alcaptonuria).

Certain drugs used for the treatment of urinary tract infections produce bright green or blue colors which usually are so obvious that the patient is aware that this is an artifactual presentation. Also, around the holiday season, one frequently encounters bright red urine samples resulting from the ingestion of certain food dyes (13). The suspicion that an organic dye is involved can frequently be confirmed by simply switching from an alkaline to an acid pH which often produces striking color changes. As a general rule, most compounds of biological importance such as hemoglobin, myoglobin, etc., do not change dramatically with shifts in pH.

The list of compounds producing odors, crystals and colors in urine is one that constantly grows and will be added to from time to time. Past experience indicates that the keen observer will not only recognize patients with known disorders but will also discover new and as yet unrecognized disorders by this, the simplest of all screening tests.

References

1. Cone, T. E., Jr.: Diagnosis and treatment. Some diseases, syndromes, and conditions associated with an unusual odor, Pediatrics 41:993-995, 1968.

2. Body odor and metabolic defects, Nutrition Abstr. & Rev. 26:107-111, 1968.

3. Buist, N. R. M.: Set of simple side-room urine tests for detection of inborn errors of metabolism, Brit. M. J. 2:745-749, 1968.

4. Snyderman, S. E.: Diagnosis of metabolic disease, Pediat. Clin. North America 18:199-208, 1971.

5. Menkes, J. H., Hurst, P. L., and Craig, J. M.: A new syndrome. Progressive familial infantile cerebral dysfunction associated with an unusual urinary substance, Pediatrics 14:462-466, 1954.

6. Tanaka, K., Budd, M. A., Efron, M. L., and Isselbacher, K. J.: Isovaleric acidemia. A new genetic defect of leucine metabolism, Proc. Nat. Acad. Sc. 56:236-242, 1966.

7. Budd, M. A., Tanaka, K., Holmes, L. B., Efron, M. L., Crawford, J. D., and Isselbacher, K. J.: Isovaleric acidemia, New England J. Med. 277:321-327, 1967.

8. Sidbury, J. B., Jr., Harlan, W. R., and Wittels, B.: Description of an apparent inborn error of short-chain fatty acid metabolism, Am. J. Dis. Child. 104:531, 1962.

9. Sidbury, J. B., Jr., Smith, E. K., and Harlan, W.: An inborn error of short-chain fatty acid metabolism, J. Pediat. 70:8-15, 1967.

10. Huguley, C. M., Jr., Bain, J. A., Rivers, S. L., and Scoggins, R. B.: Refractory megaloblastic anemia associated with excretion of orotic acid, Blood 14:615-634, 1959.

11. Cone, T. E., Jr.: Diagnosis and treatment. Some syndromes, diseases, and conditions associated with abnormal coloration of the urine or diaper, Pediatrics 41:654-658, 1968.

12. Drummond, K. N., Michael, A. F., Ulstrom, R. A., and Good, R. A.: The blue diaper syndrome. Familial hypercalcemia with nephrocalcinosis and indicanuria, Am. J. Med. 37:928-948, 1964.

13. Levin, S.: Red Urine. The Monday morning disorder of children, Pediatrics 36:134-135, 1965.

14. Smith, A. J., and Strang, L. B.: An inborn error of metabolism with the urinary excretion of a-hydroxybutyric acid and phenylpyruvic acid, Arch. Dis. Childhood 33:109-113, 1958.

15. Hooft, C., Timmermans, J., Snoeck, J., Antener, I., Oyaert, W., and Hende, C. H.: Methionine malabsorption in a mentally defective child, Lancet 2:20, 1964.

16. Perry, T. L., Hardwick, D. F., Dixon, G. H., Dolman, C. L., and Hansen, S.: Hypermethioninemia. A metabolic disorder associated with cirrhosis, islet cell hyperplasia, and renal tubular degeneration, Pediatrics 36: 236-250, 1965.

17. Humbert, J. R., Hammond, K. B., Hathaway, W. E., Marcoux, J., and O'Brien, D.: The stale-fish syndrome. A new metabolic disorder associated with trimethylaminuria, Pediat. Res. 5:395, 1971.

18. Jepson, J. B.: Hartnup Disease in Stanbury, J. B., Wyngaarden, J. B., and Fredrickson, D. S. (eds.): *The Metabolic Basis of Inherited Disease* (3rd ed.; New York: McGraw-Hill Book Company, Inc., 1972).

CHAPTER 3

The Ferric Chloride Test

One of the first laboratory tests to find wide application in the diagnostic evaluation of patients with mental deficiency was the ferric chloride test on urine. This method, which had been accepted as a qualitative test for aromatic hydroxyl groups (1), was introduced as a medical screening tool by Fölling when he described the characteristic green color of ferric chloride with urine obtained from patients with phenylpyruvic oligophrenia (phenylketonuria) (2,3).

Phenylketonuria

This inherited disorder, which is now known to be the result of a failure of the liver enzyme phenylalanine hydroxylase to convert phenylalanine to tyrosine (4), is characterized biochemically by an excessive accumulation of phenylalanine and its metabolites. It is the marked elevation in the concentration of the urinary metabolites (chiefly phenylpyruvic acid) that is responsible for the formation of the characteristic color in the presence of ferric chloride (5).

The increased concentration of the metabolites in urine is secondary to the elevation of blood phenylalanine which on transamination yields phenylpyruvic acid. The ferric chloride test does not, however, yield a positive result in many patients until the blood phenylalanine is markedly elevated (6), usually around 15 mg/100 ml. There is, therefore, in many phenylketonuric patients a period of several days to weeks before a positive result will be obtained with this screening test (6). As a result of this phenomenon the direct measurement of blood phenylalanine has become the standard method for the detection of patients with phenylketonuria in the newborn population (7).

The ferric chloride test still remains, nevertheless, an important screening method in the diagnostic work-up of families with mental retardation. This is, in part, due to the fact that phenylketonuric patients born prior to the introduction of newborn screening programs around 1965 may still remain undiagnosed. In addition, either the ferric chloride test or the Guthrie "Inhibition Test" can be used as a screening test for phenylketonuric women who are at an extremely high risk for producing offspring with both intrauterine and postnatal growth retardation, mental retardation and/or structural abnormalities (8).

Histidinemia

The ferric chloride test also retains its value as a screening tool, since characteristic colors are also formed in the presence of urine from patients with a number of other biochemical disorders associated with developmental problems (Table 3-1). One of these disorders is histidinemia. In this disorder, an inherited deficiency of the enzyme histidase results in a block in the conversion of histidine to urocanic acid (9,10). This in turn is associated with a marked

TABLE 3-1.—Inborn Errors of Metabolism Which May Be Associated With Positive Ferric Chloride Reactions

Disorder	Major Compound(s) In Urine	References
Phenylketonuria	Phenylpyruvic acid	2,3
Histidinemia	Imidazolepyruvic acid	11
Maple syrup urine disease	α-Ketoisovaleric α-Ketoisocaproic α-Keto-β-methylvaleric	20,22
Hereditary tyrosinemia with hepatorenal disease	p-Hydroxyphenyllactic acid p-Hydroxyphenylacetic acid p-Hydroxyphenylpyruvic acid	13,15
Hereditary tyrosinemia without hepatorenal disease	p-Hydroxyphenylpyruvic acid p-Hydroxyphenyllactic acid p-Hydroxyphenylacetic acid	43 44
Tyrosinosis (Mede's Disease)	p-Hydroxyphenylpyruvic acid	19
Alcaptonuria	Homogentisic acid	20,21
Oasthouse disease	α-Hydroxybutyric acid phenylpyruvic acid	23
Formiminotransferase— deficiency syndrome ($\pm$)	5-Amino-4-imidazol-carboxamide	18

increase in the urinary excretion of histidine and at least one of its metabolites, imidazole-pyruvic acid. This latter compound yields a positive blue-green color in the presence of ferric chloride (11). While this color is similar to that obtained from patients with phenylketonuria (PKU), it differs in that it is quite stable, while the color formed by phenylpyruvic acid in urine from PKU patients fades within a few hours.

Due to the fact that the enzyme histidine transaminase, which converts histidine to imidazolepyruvic acid, appears to be inactive for the first several weeks or months of life, there may be a delay in the accumulation of the urinary imidazolepyruvic acid in newborns affected with histidinemia. For this reason the ferric chloride test may yield a negative result in patients with histidinemia during early life and is, therefore, not an effective screening method for the newborn population (12).

Tyrosinemia

Another compound of clinical importance which yields a positive result (transient green or blue) in the presence of ferric chloride under appropriate conditions is p-hydroxyphenyl-pyruvic acid (13). This compound is a normal intermediate in the pathway for tyrosine metab-olism. A number of clinical conditions have been described which involve an increase in this and certain closely related compounds such as p-hydroxyphenyllactic acid and p-hydroxy-phenylacetic acid. Most of these conditions appear to be associated with an impairment in the liver p-hydroxyphenylpyruvic acid oxidase activity which is required to convert p-hydroxy-phenylpyruvic acid to homogentisic acid.

The reduction of this enzyme activity may be of a transient nature as is often found in premature infants (14), of a permanent nature as is found in the inherited condition tyrosinemia (15), or of a secondary nature as is found in severe liver damage due to any of a number of causes (16). As the p-hydroxyphenylpyruvic acid is, for the most part, reduced to p-hydroxyphenyllactic acid, this compound, not the keto acid, is the main tyrosyl product in the urine of patients with the above disorders (17). In keeping with the observation that only the p-hydroxyphenylpyruvic acid reacts with ferric chloride while the p-hydroxyphenyllactic acid does not, urine of patients with these disorders may, in many cases, yield a negative result with the ferric chloride test. It is thus recommended that the nitrosonaphthol test described in Chapter 5 be used as the preferred method for detecting patients excreting increased amounts of p-hydroxyphenyllactic acid in the urine.

In contrast, however, a single patient with tyrosinosis described in detail by Medes was found to excrete very high concentrations of the p-hydroxyphenylpyruvic acid and little, if any, of the reduced compound (19). The exact nature of the enzyme defect in this disorder is still unclear (17); however, additional cases, if found, should yield a positive ferric chloride test result.

Alcaptonuria

There is also an inherited defect involving the next enzymatic conversion in this same metabolic pathway. The absence of this enzyme, homogentisic acid oxidase, which is required to convert homogentisic acid to maleylacetoacetic acid, results in a marked increase in the excretion of homogentisic acid, which again is responsible, upon the addition of small amounts of ferric chloride, for a transient green or blue color (20,21). This condition is known as alcaptonuria as the urine darkens on standing, particularly when alkaline.

Maple Syrup Urine Disease

The ferric chloride test also yields a characteristic, although quite different, reaction with the urine of patients with the inherited disorder of maple syrup urine disease. In this disorder an impairment in the oxidative decarboxylation of the a-keto acids—a-ketoisovaleric acid, a-ketoisocaproic acid and a-keto-β-methylvaleric acid—results in a marked increase in the urinary excretion of these keto compounds and their related amino acids—valine, leucine and isoleucine. In the presence of excessive concentrations of these keto acids, the addition of ferric chloride yields a color which might be described as grey with a greenish tinge (20) and dark green (22).

Oasthouse Disease

A persistent green color in the presence of ferric chloride is also found in at least one additional inherited disorder of metabolism. This is the so-called Oasthouse disease of Smith and Strang (23) in which large amounts of a-hydroxybutyric acid and phenylpyruvic acid are excreted. It is of importance to note that while the reaction of a-hydroxybutyric acid in the presence of ferric chloride yields a purple color which fades to a red-brown color, the urine of this patient yielded a green color, presumably due to excessive amounts of phenylpyruvic acid (23).

Noninherited Disorders

In addition to the inborn errors of metabolism which yield a colored product with ferric chloride, there are many noninherited disorders which also can yield positive results. These range from disorders associated with excessive excretion of bilirubin (24) to a variety of tumors such as melanoma and pheochromocytoma (22) (see Table 3-2).

TABLE 3-2.—Nonhereditary Disorders Which May Be Associated With a Positive Ferric Chloride Reaction

Disorder	Major Reactive Compound(s) in Urine	Color Reaction	References
Transient tyrosinemia	p-Hydroxyphenyllactic acid p-Hydroxyphenylacetic acid p-Hydroxyphenylpyruvic acid	Transient blue or green ($\pm$)	(35, 36) (20)
Cirrhosis of the liver	Same as above	Transient blue or green ($\pm$)	(22)
Icterus	Bilirubin (conjugated)	Green—stable	(24, 37)
Diabetic acidosis	Acetoacetic	Purple or cherry red	(38)
Melanoma	Melanin	Gray ppt.—turning black	(13, 38)
Pheochromocytoma	Catecholamines	Blue-green	(21, 22)
Carcinoid	5-Hydroxyindolacetic	Blue-green or dark green	(22)

Medications

It has also been found that urine from persons ingesting a wide variety of medications yield, in many cases, colored reaction products with ferric chloride (Table 3-3). The most commonly encountered compounds in this group are the metabolites of aspirin which yield a deep purple color. In addition to the possibility of mistaking this reaction with the clinically important reactions, there is also the possibility that the deep color produced by the salicylate could "mask" a positive reaction for a clinically important disease (25).

Another reaction which occurs with less frequency, but which potentially might also cause confusion, is that found with the urine of patients under treatment with L-Dopa. The potential for confusion results from the fact that urine samples from these patients yield, upon the addition of ferric chloride, a positive green color, which vanishes and then reappears as a stable emerald green color (26). While evidence has been presented which indicates that the L-Dopa metabolite, dopylacetic acid, is responsible for this reaction, the green color of this reaction might easily be interpreted as a positive test for any one of several of the inherited metabolic disorders already discussed (see Tables 3-1 and 3-4).

Other medications and/or their related metabolites which have been reported to yield positive results with ferric chloride are listed in Table 3-3. Because of the wide use of many of these

TABLE 3-3.—Urinary Ferric Chloride Reactions in Patients Ingesting Various Drugs

Drugs	Color Reaction	References
Salicylates	Purple	(41, 20)
Phenothiazine derivatives	Purple	(39, 40, 41)
Antipyrine	Red	(22)
Isonicotinic acid hydrazide (isoniazid)	Yellowish green to brown-violet	(22)
L-Dopa metabolites	Green	(26)
Nystaform-HC ointment (iodochlorhydroxyquin)	Blue-green	(42)

TABLE 3-4.—Comparison of Results of Various Methods for Performing the Ferric Chloride Reactions With Compounds Excreted, in Excess, in a Variety of Inborn Errors of Metabolism*

Compound*	Method of Renuart (34)	Method of Meulemans (30)	Method of Perry (32)	Modified Method of Perry	Phenistix
Phenylpyruvic acid	Dark green	Dark green	Dark green	Dark green	Green
Imidazolepyruvic acid	Dark green	Blue	Blue	Purple	Gray
a-Ketoisocaproic acid	Purple	Purple	Purple	Purple	Negative
a-Ketoisovaleric acid	Negative	Negative	Negative	Negative	Negative
p-Hydroxyphenyllactic acid	Negative	Negative	Negative	Negative	Negative
p-Hydroxyphenylacetic acid	Negative	Negative	Negative	Negative	Negative
p-Hydroxyphenylpyruvic acid	Transient green	Transient green	Negative	Light green	Negative
Homogentisic acid	Transient blue	Negative	Negative	Transient blue	Negative

*Test compound: 100 mg/100 ml of distilled water.

compounds one should, if possible, be aware of the recent medication history of any patients whose urine is tested with the ferric chloride test.

As is evident from Tables 3-1 and 3-2, ferric chloride in solution yields a variety of colors with a large number of organic compounds. One should be aware that this list of reactive compounds is not limited to the relatively few compounds already discussed, but actually includes a long list of phenols, enols, oximes, and some carboxylic acids (1,27). We consider the lack of specificity a distinct advantage for a screening test; however, it requires that additional tests be utilized to confirm a specific diagnosis.

While the exact nature of the chemical reaction leading to the production of color by ferric chloride appears to be poorly understood, some information is available regarding the effects of a number of commonly encountered factors upon the reaction. These factors include temperature, solvent, reactant concentrations, and acidity. While most of the available information is based on studies carried out with phenylpyruvic acid, it is likely that many of the same factors influence the reaction with many of the other medically important organic compounds.

Stability of Samples

From a practical standpoint, the condition of the urine upon arrival in the laboratory is of major importance. Henry, for example, quotes reports that phenylpyruvic acid decomposes in an alkaline solution; therefore, if the urine is alkaline or becomes alkaline through bacterial action, there is a danger that the urine will yield a "false negative" result (13). In keeping with this observation Paine reports that although most samples from PKU patients will remain positive for days, especially if refrigerated, he has encountered a case where an initially positive urine sample became negative within 1½ hours (28). Methods of preservation suggested by various authors include refrigeration plus thymol crystals (5) and refrigeration plus 0.5% H_2SO_4 (28).

pH Effects

In addition to the decomposition problem, Rupe and Free (29) also point out that an alkaline urine containing large amounts of phenylpyruvic acid could yield a negative result as a result of the precipitation of the reactive ferric ions as ferric hydroxide. Centerwall et al. (21) report that in spite of the fact that many textbooks state that urine should be acidified before performing the ferric chloride test, this is not necessary if sufficient ferric chloride is added to the sample being tested. If one, for example, performs the test using 3-5 drops of a 10% ferric chloride solution with 1 ml of urine, the acidity of the ferric chloride solution (pH 1.8) counteracts any excess of basic ions present in the urine. If, on the other hand, one uses a 5% ferric chloride solution and a 5-ml sample of urine, it might happen that ferric hydroxide would form, thus yielding poor results.

It should also be emphasized that the ferric ions can also be lost as ferric hydroxide due to leaching out of alkali from glass bottles in which the ferric chloride solution is stored. For this reason we recommend that the stock solution be stored in a polyethylene or stained glass bottle in a refrigerator. A smaller "working solution" can then be stored at room temperature for day to day use.

Interference by Phosphates

It has also been shown that the sensitivity of the ferric chloride test is influenced by phosphates which are present in varying amounts in urine. Rupe and Free, for example, found that the sensitivity was markedly reduced by the addition of NaH_2PO_4 to final levels which are often encountered in routine urine samples (29). In an attempt to overcome this difficulty, Meulemans uses a mixture of magnesium chloride, ammonium chloride and ammonia to precipitate the phosphates and thus improve the sensitivity of the test (30).

Transient Nature of the Reaction

Another troublesome aspect of the ferric chloride reaction is the transitory nature of the color reaction with many of the organic compounds of medical importance including phenylpyruvic acid. A number of modifications have been introduced to overcome this difficulty. For example, Saifer and Harris have studied various aspects of this problem in an attempt to develop a satisfactory quantitative assay for urinary phenylpyruvic acid (31). This group, as well as others, found that the rate of color decay can be reduced by ferrous ions, absence of light and low temperature. One of these modifications, the addition of ferrous ammonium sulfate, is employed in the qualitative test of Perry et al. (32).

Paper Strip Technique

In addition to the liquid ferric chloride tests, the Ames Company of Elkhart, Indiana, markets a paper strip (Phenistix) which can be used as a screening tool for many of the same compounds already discussed.

The paper strip screening technique was first developed by Baird who impregnated a portion of a paper stick with ferric chloride and glacial acetic acid (33). The Phenistix strips are also impregnated with ferric chloride; however, additional modifications are included in these strips. These include the addition of the buffer cyclohexyl-sulfamic acid which provides the optimal pH for color development and magnesium ions to reduce the interference by urine phosphates (29). When this strip is moistened with urine containing phenylpyruvic acid, a gray to blue color is formed. As with the various liquid tests, however, this test also reacts with a wide variety of compounds and care must be taken in the interpretation of positive results.

Variable Results

Yet another troublesome aspect of the ferric chloride test is that color reactions which are described as being characteristic are in fact, in many cases, only characteristic if the test is performed in a particular manner. Several of the compounds of clinical importance yield a variety of results, including negative, by different modifications of the test (see Table 3-4). In this regard our own experience suggests that the most important variable may be the relative concentrations of the reactive compounds and the ferric chloride. Specifically the "p-hydroxy-phenyl-compounds," homogenistic acid and acetoacetic acid, which yield positive results if small amounts of the ferric chloride solution is added to a standard solution, give negative tests results by the standard methods of Meulemans (30) and Perry et al. (32).

The Ferric Chloride Test

Method of Renuart (34)

Reagents:

1. Ferric chloride 10%: Dissolve 10 Gm of $FeCl_3$ in a small amount of water and dilute to a total of 100 ml with water.

Procedure:

1. Place 1-2 ml of *fresh* urine in a clean test tube.

2. Add 10% ferric chloride one drop at a time.

3. Observe for color.

Results:

1. See Tables 3-1 — 3-4 for interpretation.

The Ferric Chloride Test

Method of Meulemans (30)

Reagents:

1. Demasking reagent: Dissolve 1.1 Gm of magnesium chloride and 1.4 Gm of ammonium chloride in distilled water. Add 2.25 ml of ammonium hydroxide and dilute to a final volume of 100 ml with distilled water.

2. Ferric chloride solution, 10%: Dissolve 5 Gm of $FeCl_3$ in a small amount of distilled water and dilute to a final volume of 50 ml with water.

3. Hydrochloric acid, 10%: Add 27.7 ml of concentrated HCl to distilled water and dilute to a final volume of 100 ml.

Method:

1. Add 1 ml of the "demasking" reagent to 4 ml of urine. Mix and allow to stand for 5 minutes.

2. Filter.

3. Acidify filtrate with two drops of the 10% hydrochloric acid. Check pH with pH paper. If not acid, add more 10% hydrochloric acid.

4. Add 2 drops of the 10% ferric chloride solution.

5. Observe color. Allow to stand for 30 minutes and observe for color a second time.

Results:

1. See Tables 3-1—3-4 for interpretation.

The Ferric Chloride Test

Method of Perry et al. (32) (This is the method routinely used by the authors.)

Reagents:

1. Hydrochloric acid, 1.0N stock: Add 8.33 ml of concentrated hydrochloric acid to about 75 ml of distilled water and dilute to a total of 100 ml with distilled water.
2. Hydrochloric acid, 0.02N working reagent: Add 2 ml of the 1.0N HCl stock solution to water and dilute to a total volume of 100 ml with water.
3. Ferric chloride reagent: Dissolve 1.0 Gm of ferric chloride and 1.0 Gm of ferrous ammonium sulfate [$Fe(NH_4)_2(SO_4)_2 \cdot 6H_2O$] in 0.02N hydrochloric acid and dilute to a total volume of 100 ml with 0.02N HCl. This reagent can be stored at room temperature.

Procedure:

1. Place 1 ml of ferric chloride reagent in a clean test tube.
2. Add 10 drops (0.5 ml) urine and mix well by shaking.
3. Observe for color.

Results:

1. See tables for interpretation.
2. It should be noted that "p-hydroxyphenyl-compounds" and homogentisic acid will not yield a transient blue color by this procedure. A positive test for these compounds in concentrations above 10 mg/100 ml can, however, be obtained with the following modification:
    a. Place 1 ml of urine in a test tube.
    b. Add 1-2 drops of Perry reagent.
    c. Observe for "flash of blue" color.

Ferric Chloride Test Using PHENISTIX

Reagents:

1. Phenistix, Ames Company, Inc., Elkhart, Indiana.

Method:

1. Dip test end of strip in urine and remove immediately or moisten by pressing against wet diaper or filter paper.
2. Compare color of dipped end with color chart at ½ minute. (Color chart on bottle label.)

Results:

1. See Table 3-4.

References

1. Wesp, E. F., and Brode, W. R.: The absorption spectra of ferric compounds. I. The ferric chloride-phenol reaction, J. Am. Chem. Soc. 56:1037-1042, 1934.

2. Fölling, A.: Über Ausscheidung von Phenylbrenztraubensäure in den Harn als Stoffwechsel-anomalie in Verbindung mit Imbezillität, Ztschr. Physiol. Chem. 227:169-176, 1935.

3. Fölling, A.: Utskillelse av fenylpyrodruesyre i urinen som Stoffskifteanomali i forbindelse med imbecillitet, Nord. med. tidskr. 8:1054-1059, 1934.

4. Jervis, G. A.: Phenylpyruvic oligophrenia deficiency of phenylalanine oxidizing system, Proc. Soc. Exper. Biol. & Med. 82:514-515, 1953.

5. Penrose, L., and Quastel, J. H.: XXXVIII. Metabolic studies in phenylketonuria, Biochem. J. 31:266-274, 1937.

6. Efron, M. L., and MacCready, R. A.: Phenylketonuric families, J.A.M.A. 191:391-392, 1965.

7. Guthrie, R., and Susi, A.: A simple phenylalanine method for detecting phenylketonuria in large populations of newborn infants, Pediatrics 32:338-343, 1963.

8. Howell, R. R., and Stevenson, R. E.: The offspring of phenylketonuric women, Soc. Biol. 18:19-33, 1971.

9. Ghadimi, H., Partington, M. W., and Hunter, A.: A familial disturbance of histidine metabolism, New England J. Med. 265:221-224, 1961.

10. LaDu, B. N., Howell, R. R., Jacoby, G. A., Seegmiller, J. E., and Zannoni, V. G.: The enzymatic defect in histidinemia, Biochem. Biophys. Res. Commun. 7:398-402, 1962.

11. Auerbach, V. H., DiGeorge, A. M., Baldridge, R. C., Tourtellotte, C. D., and Brigham, M. P.: Histidinemia. A deficiency in liver histidase resulting in the urinary excretion of histidine and of imidazolepyruvic acid, Clin. Res. 9:334, 1961.

12. Levy, H. L., Madigan, P. M., and Peneva, P.: Evidence for delayed histidine transamination in neonates with histidinemia, Pediatrics 47:128-132, 1971.

13. Henry, R. J.: *Clinical Chemistry: Principles and Techniques* (New York: Paul B. Hoeber, Inc., Harper & Row, Publishers, 1964).

14. Kretchmer, N., Levine, S. Z., and McNamara, H.: The in vitro metabolism of tyrosine and its intermediates in the liver of the premature infant, Am. J. Dis. Child. 93:19, 1957.

15. Gentz, J., Jagenburg, R., and Zetterström, R.: Tyrosinemia. An inborn error of tyrosine metabolism with cirrhosis of the liver and multiple renal tubular defects (de-Toni-Debre-Franconi syndrome), J. Pediat. 66:670-696, 1965.

16. Levine, R. J., and Conn, H. O.: Tyrosine metabolism in patients with liver disease, J. Clin. Invest. 46:2012-2020, 1967.

17. LaDu, B. N.: Tyrosinosis in Stanbury, J. B., Wyngaarden, J. B., and Fredrickson, D. S. (eds.): *The Metabolic Basis of Inherited Disease* (2nd ed.; New York: McGraw-Hill Book Company, Inc., 1966).

18. Arakawa, Ts., Ohara, K., Takahashi, Y., Ogasawara, J., Hayashi, T., Chiba, R., Wada, Y., Tada, K., Mizuno, T., Okamura, T., and Yoshida, T.: Formiminotransferase deficiency syndrome. A new inborn error of folic acid metabolism, Ann. paediat. 205:1-11, 1965.

19. Medes, G.: A new error of tyrosine metabolism: Tyrosinosis. The intermediary metabolism of tyrosine and phenylalanine, Biochem. J. 26:917-940, 1932.

20. Gibbs, N. K., and Woolf, L. I.: Tests for phenylketonuria, Brit. M. J. 2:532-535, 1959.

21. Centerwall, W. R., Chinnock, R. F., and Pusavat, A.: Phenylketonuria. Screening programs and testing methods, Am. J. Pub. Health 50:1667-1677, 1960.

22. Hyanek, S.: Zur methodik des Ferrichloridtests (Fölling-Reaktion) und seiner Auswertung, Z. Med. Labortech. 9:365-370, 1968.

23. Smith, A. J., and Strang, L. B.: An inborn error of metabolism with the urinary excretion of a-hydroxybutyric acid and phenylpyruvic acid., Arch. Dis. Childhood 33:109-13, 1958.

24. Jervis, G. A.: Phenylpyruvic oligophrenia (phenylketonuria), Res. Publ., A. Nerv. & Ment. Dis. 33:259-282, 1954.

25. Heffelfinger, J. C., and LaGro, R. J.: False negative urine tests in surveys for phenylketonuria, Pediatrics 25:1086-1087, 1960.

26. Iversen, P. F., Closs, K., and Wad, N.: Darkening of urine from patients treated with L-Dopa, Clin. chim. acta 32:137-139, 1971.

27. Soloway, S., and Wilen, S. H.: Improved ferric chloride test for phenols, An. Chem. 24:979-983, 1952.

28. Paine, R. S.: Testing urine for phenylpyruvic acid, Pediatrics 22:1027, 1958.

29. Rupe, C. O., and Free, A. H.: An improved test for phenylketonuria, Clin. Chem. 5:405-413, 1959.

30. Meulemans, O.: The ferric chloride test for phenylpyruvic acid in urine, Clin. chim. acta 5:152-153, 1960.

31. Saifer, A., and Harris, A. F.: Studies on the photometric determination of phenylpyruvic acid in urine, Clin. Chem. 5:203-217, 1959.

32. Perry, T. L., Hansen, S., and MacDougall, L.: Urinary screening tests in the prevention of mental deficiency, Canad. M. A. J. 95:89-95, 1966.

33. Baird III, H. W.: A reliable paper strip method for the detection of phenylketonuria, J. Pediat. 52:715-718, 1958.

34. Renuart, A. W.: Screening for inborn errors of metabolism associated with mental deficiency or neurologic disorders or both, New England J. Med. 274:384-387, 1966.

35. Gilbert, E. F.: An evaluation of some laboratory diagnostic tests in phenylketonuria, Clin. Proc. Child. Hosp., Washington 18:328-332, 1962.

36. Thomas, G. H.: Unpublished results.

37. Wright, S. W.: Phenylketonuria, J.A.M.A. 165:2079-2083, 1957.

38. Zaccaria, A., Jr.: Ferric chloride test on urine, J.A.M.A. 165:200-201, 1957.

39. Forrest, F. M., Forrest, I. S., and Mason, A. S.: Review of rapid urine tests for phenothiazine and related drugs. Am. J. Psychiat. 118:300-307, 1961.

40. Pollack, B.: The Validity of Forrest reagent test for the detection of chlorpromazine or other phenothiazines in urine, Am. J. Psychiat. 115:77-78, 1958.

41. Nellhaus, G.: Clinical use of Phenistix reagent strip method of testing urine samples, J.A.M.A. 170:1052-1053, 1959.

42. Kane, R. B., Kratka, H. B., and Katz, M.: Sources of error in testing for PKU, Pediatrics 41:1146, 1968.

43. Wadman, S. K., Van Sprang, F. J., Maas, J. W., and Ketting, D.: An exceptional case of tyrosinosis, J. Ment. Defic. Res. 12:269-281, 1968.

44. Holston, J. L., Levy, H. L., Tomlin, G. A., Atkins, R. J., Patton, T. H., and Hosty, T. S.: Tyrosinosis. A patient without liver or renal disease, Pediatrics 48:393-400, 1971.

CHAPTER 4

Ketoaciduria: Dinitrophenylhydrazine and Thin Layer Chromatography Procedures

It is now well established that there are several inborn errors of metabolism associated with a marked increase in the excretion of one or more organic keto acids in the urine (1,2) (Table 4-1). This group of carbonyl compounds includes both aromatic a-keto acids as are found, for example, in phenylketonuria (3) and tyrosinosis (4), and aliphatic a-keto acids as are found in maple syrup urine disease (5).

Chemical Reaction

As each of the above disorders are reviewed in Chapters 3 and 5, this discussion will be limited to the reaction of these compounds with the laboratory reagent 2,4-dinitrophenylhydrazine (DNPH). This compound has been utilized for many years as a screening reagent as it forms relatively insoluble crystalline hydrazones with a variety of aliphatic, cyclic and aromatic carbonyl compounds of medical importance. This reaction proceeds in the following manner:

$$\begin{matrix} R_1 \\ \quad \diagdown \\ \quad\quad C = 0 \\ \quad \diagup \\ R_2 \end{matrix} + NH_2 \cdot NH \cdot C_6H_3(NO_2)_2 \longrightarrow \begin{matrix} R_1 \\ \quad \diagdown \\ \quad\quad C = N \cdot NH \cdot C_6H_3(NO_2)_2 \\ \quad \diagup \\ R_2 \end{matrix} + H_2O$$

As this reaction results in an insoluble product, a yellow to yellow-white precipitate forms upon the addition of this compound to urine samples containing excessive amounts of a variety of keto compounds (7,14). As this result is easily distinguished from that obtained from normal urine samples in which no precipitate forms, it can be used as a rapid screening test for patients with an assortment of pathological conditions associated with increased keto acid concentrations (14,15).

Normal Findings

It should perhaps be noted that urine from normal controls is not completely devoid of a-keto acids. With sufficiently sensitive methods one can demonstrate the presence of several a-keto acids in these samples, the major ones being pyruvic acid and a-ketoglutaric acid (1,8). As each of these compounds is normally present in relatively small amounts, however, the DNPH screening test routinely yields a negative result with urine samples from normal controls.

One possible exception to this generalization might be encountered in urine samples from newborn infants. Snyderman, for example, reports that the DNPH test is usually moderately positive during the first few days of life as a result of the increased excretion of pyruvic acid, a-ketoglutaric acid and acetoacetic acid (16).

TABLE 4-1.–Inherited Disorders Associated With an Increased Excretion of Keto Acids

Disorder	Major Compounds	References
Phenylketonuria	Phenylpyruvic acid	3,7
Maple syrup urine disease	a-Keto isovaleric acid a-Keto isocaproic acid a-Keto-γ-methylbutyric acid	5,8
Oasthouse disease	a-Hydroxybutyric acid† Phenylpyruvic acid	6
Tyrosinosis (Medes Disease)	p-Hydroxyphenylpyruvic acid	4
Tyrosinemia*	p-Hydroxyphenylpyruvic acid	9,10
Pyruvic acidemia	Pyruvic acid	11,12
Hypermethioninemia	p-Hydroxyphenylpyruvic acid a-Keto-γ-methiobutyric acid Phenylpyruvic acid	13
Histidinemia	Imidazolepyruvic acid†	2

*See Chapters 3 and 5 for discussion.
†This compound does not react with the DNPH test.

Ketone Bodies

One must also keep in mind that urine samples containing "ketone bodies" will also yield a positive result with the DNPH test. The ketone bodies, acetoacetic acid, β-hydroxybutyric acid and acetone are generally found in excessive amounts when the major source of metabolic energy shifts from carbohydrate to fat.

The increase in fat utilization leads first to an excessive formation and accumulation of acetoacetic acid. This compound can then be converted to acetone and β-hydroxybutyric acid. While all three of these compounds can be excreted in urine, it appears that acetoacetic acid is primarily responsible for the positive DNPH result. As the production of these compounds can occur in any metabolic disturbance in which the carbohydrates are improperly utilized, one will obtain a positive result in urine samples from patients suffering from any one of a large number of specific and nonspecific disorders associated with the production of ketone bodies.

Our experience indicates that the use of Acetest* tablets, which react with acetone and acetoacetic acid, while being unreactive to the a-keto acids, is a useful means of distinguishing between urine samples containing excessive amounts of the a-keto acids and those containing primarily ketone bodies.

While this has been found to be a useful approach, it is also possible to form misleading conclusions as certain primary a-keto acid disorders will, on occasion, be associated with the excretion of ketone bodies, i.e., maple syrup urine disease.

It must also be remembered that there are a number of genetic disorders which, while not being associated with increased a-keto acid excretion, are associated with the excretion of

---

*Acetest is the registered trademark of the Ames Laboratory, Inc., Elkhart, Indiana.

TABLE 4-2.—Inherited Disorders Associated With
Ketosis*

Ketotic hyperglycinaemia

Isovaleric acidaemia

Glycogen storage disease, type I

Methylmalonic aciduria

Diabetes

Fructose 1-6, diphosphatase deficiency

Lactic acid acidosis

Butyric/hexanoic acidemia

*"Ketone bodies" will yield a positive DNPH
test.

ketone bodies (14). Some of the disorders included in this group are shown in Table 4-2. It is, therefore, important that this laboratory finding not be dismissed lightly as it might serve as an important clue for the recognition of these disorders.

Chromatographic Techniques

As can be seen from the above discussion the DNPH test, while a very useful screening tool, is nevertheless based on a rather nonspecific reaction associated with a high percentage of positive results. As many of these positive findings are due to nongenetic causes, it is often important to employ additional laboratory procedures which are capable of distinguishing the various reactive compounds (17).

The technique of chromatography has been found to be a very satisfactory method for this purpose. The various techniques employed for the study of the keto acids in the urine and other body fluids have included column chromatography (18), paper chromatography (1,19,20), and more recently thin layer chromatography (8,21). The ease and rapidity of this layer chromatography has caused it currently to be the most widely used.

Each of these procedures is based on the formation, extraction and separation of the 2,4-dinitrophenylhydrazone derivatives of α-keto compounds. While it is true that these techniques cannot be employed as routine screening procedures for a large number of urine samples, they are sufficiently simple to allow for the detailed study of urine samples which yield inconclusive results by the simple screening tests already discussed.

Qualitative Dinitrophenylhydrazine (DNPH) Test for Keto Compounds

Method of Penrose and Quastel (7)

Reagents:
1. Hydrochloric acid, 1N: Add 8.33 ml of concentrated HCl to about 75 ml of distilled water and dilute to a total volume of 100 ml with distilled water.

2. Dinitrophenylhydrazine reagent: Dissolve 0.7 Gm of 2,4-dinitrophenylhydrazine (DNPH) in 250 ml of 1N hydrochloric acid by warming. Cool and store in a dark bottle. If supernatant is not clear yellow solution, filter the solution before storing (7).

Procedure:

1. Filter urine.

2. Add 1 ml of DNPH reagent to 1 ml of filtered urine in a small test tube.

Results:

1. Immediately upon the addition of the DNPH reagent, the mixture is a clear, pale yellow-orange solution. It will remain this way if the test is negative.

2. The test is positive when a yellow to yellow-white turbidity or precipitate forms within 5 minutes. Penrose and Quastel describe a positive result as the almost immediate formation of a yellow opalescence or yellow precipitate (7).

Notes:

1. Various procedures employed throughout the years have used concentrations of DNPH of 0.1% (5,15,22), 0.28% (7,23) and 0.4% (24). As the 0.4% solution is supersaturated it results, after cooling and filtration, in about a 0.3% solution (24).

2. Some workers employ 1N HCl (7,23,24) while others use 2N HCl (5,15,22).

3. This test is reported to be able to detect concentrations of phenylpyruvic acids as low as 5-10 mg/100 ml. (25).

Thin Layer Chromatography of a-Keto Acid 2,4-Dinitrophenylhydrazones

Modification of References 1, 8 and 20

Reagents:

1. Hydrochloric acid 1N: Add 8.33 ml of concentrated HCl to about 75 ml of distilled water and dilute to a total volume of 100 ml.

2. 2,4-Dinitrophenylhydrazine, 0.28%: Dissolve 0.7 Gm of 2,4-dinitrophenylhydrazine in 250 ml of 1N hydrochloric acid by warming. If a clear solution is not obtained, filter the solution and then store in a dark brown bottle.

3. Chloroform-ethanol extraction mixture: Add 100 ml of ethanol to 400 ml of chloroform.

4. Sodium carbonate, 10%: Dissolve 10 Gm of Na_2CO_3 in water and dilute to a final volume of 100 ml.

5. Hydrochloric acid, 5N: Add 41.66 ml of concentrated HCl to water and dilute to a final volume of 100 ml with water.

6. Ammonium hydroxide, 0.25N: Add 1.66 ml of concentrated ammonium hydroxide (15N) to distilled water and dilute to a final volume of 100 ml.

7. Chromatography solvent: Prepare this solvent by mixing isoamyl alcohol with 0.25N ammonium hydroxide in a ratio of 200:10.

8. Silica Gel G, (E. Merck AG, Darmstadt, Germany). Mix 30 Gm of silica gel G with 60 ml of water by shaking vigorously in a 250-ml Erlenmeyer flask. Mix until the mixture is free of lumps. Spread the mixture on glass plates with an appropriate spreader, i.e., Desaga/Brinkmann apparatus for thin layer chromatography, and allow to air dry for

about 1/2 hour. Activate the silica gel by heating the plates in an oven at 100-110°C. for at least 1 hour.

Preparation of Urine Sample (1,20):

1. Place a 10-ml sample of the urine to be tested in a separatory funnel.
2. Add 2 ml of the DNPH solution and mix well. Allow this solution to stand for at least 30 minutes at room temperature.
3. Transfer to a separatory funnel and extract the hydrazones with three portions (15 ml each) of the chloroform-ethanol mixture. If necessary, centrifuge to "break" the emulsion. Discard the urine.
4. Combine the three chloroform-ethanol extracts and re-extract with 15 ml of the 10% sodium carbonate solution.
5. Discard the chloroform-ethanol mixture and save the sodium carbonate solution.
6. "Wash" the sodium carbonate solution containing the hydrazones with 10 ml of the chloroform-ethanol solvent.
7. Acidify the sodium carbonate solution to a pH of 2-3 by adding about 5 ml of 5N HCl. Use pH paper to determine pH.
8. Re-extract the acidified solution with successive solutions of 10, 5, and 5 ml of the chloroform-ethanol solution.
9. Save the chloroform-ethanol solution containing the keto acid derivatives.
10. Add a small amount of anhydrous sodium sulfate (Na_2SO_4), and then filter and save the clear solution.
11. Evaporate the above solution to dryness with a gentle flow of air. (Do not heat.)
12. Dissolve the residue in a final volume 0.5 ml of ethanol.

Preparation of a-Keto Acid 2,4-Dinitrophenylhydrazone Standards (8):

1. Dissolve 50 mg of the pure keto acid in 0.5 ml. of distilled water.
2. Add 1 ml of the 2,4-dinitrophenylhydrazine in 1N HCl, mix well and allow to stand at room temperature for 1 hour in a dark location.
3. Centrifuge and remove the supernatant. Discard the supernatant and save the yellow precipitate.
4. Wash the precipitate with cold distilled water. Centrifuge and again remove the supernatant. Save the yellow precipitate.
5. Dissolve the precipitate in the smallest possible amount of ethanol. Add distilled water to this solution until a precipitate forms. Centrifuge and store the pure yellow hydrazone until needed.
6. Dissolve 1 mg of the above samples in 1.0 ml absolute ethanol. Use 5 or 10 microliters (μl) of this sample for the chromatographic procedure.

Thin Layer Chromatographic Separation (8):

1. Mark one side of the chromatographic plate with small dots 2 cm apart and 2 cm from bottom with a sharp pencil.

2. Apply a total of 25 μl of the urine sample extracts to be analyzed in 5-μl amounts allowing the samples to dry between applications. This is equivalent to 0.5 ml of the original urine.

3. Apply either 5 or 10 μl of the appropriate authentic a-keto acid 2,4-dinitrophenylhydrazones.

4. Place the plate in the chromatogram tank to which had been previously added the solvent (isoamyl alcohol, 0.25N ammonium hydroxide—200:10).

5. Allow the solvent to "run" at room temperature for about 4-6 hours.

6. At the end of this time remove the plates and allow to air dry in a hood.

7. Locate the a-keto acid 2,4-dinitrophenylhydrazones by observing the plate for the intrinsic yellow color which is characteristic of these compounds.

8. Alternatively one can observe the plate under ultraviolet (U.V.) light for the presence of dark spots as the hydrazones absorb the U.V. light.

Notes:

1. A number of the compounds are divided into two isomers by this method (8,26).

2. Some representative R_f values are shown in Table 4-3 (8,26).

3. Dancis et al. report that while the absolute R_f value may vary slightly, the relative position of the various hydrazones remains the same (8).

TABLE 4-3.—R_f Values of 2-4 Dinitrophenylhydrazones of Certain Keto Acids in Isoamyl Alcohol—0.25N NH$_4$OH (20:1) on 0.3 mm Thick Silica Gel G Thin Layer Chromatography Plates*

Compound	References
Oxaloacetic acid	0
a-Keto glutaric acid	0
Acetoacetic acid	9
Pyruvic acid (isomer No. 2)	11
Phenylpyruvic acid (isomer No. 2)	22
a-Ketoisocaproic acid (isomer No. 2)	22
Pyruvic acid (isomer No. 1)	28
a-Ketoisovaleric acid	42
Phenylpyruvic acid (isomer No. 1)	46
a-Keto-β-methylvaleric acid	48
a-Ketoisocaproic acid (isomer No. 1)	48
2-4 Dinitrophenylhydrazine	73
Acetone-DNPH	84

*From Dancis et al. (8) and Dancis and Levitz (26).

References

1. Menkes, J. H.: The pattern of urinary alpha keto acids in various neurological diseases, Am. J. Dis. Child. 99: 500-506, 1960.

2. Humbel, R.: Maladies du systeme nerveux central avec elimination urinaire pathologique d'acides alpha-cetoniques, Encéphale 54:174-186, 1965.

3. Meister, A.: Phenylpyruvic oligophrenia, Pediatrics 21: 1021-1031, 1958.

4. Medes, G.: A new error of tyrosine metabolism: Tyrosinosis. The intermediary metabolism of tyrosine and phenylalanine, Biochem. J. 26:917-940, 1932.

5. Menkes, J. H.: Maple syrup disease. Isolation and identification of organic acids in the urine, Pediatrics 23:348-353, 1959.

6. Smith, A. J., and Strang, L. B.: An inborn error of metabolism with the urinary excretion of a-hydroxy-butyric acid, and phenylpyruvic acid, Arch. Dis. Childhood 33:109-113, 1958.

7. Penrose, L., and Quastel, J. H.: Metabolic studies in phenylketonuria, Biochem. J. 31:266-274, 1937.

8. Dancis, J., Hutzler, J., and Levitz, M.: Thin-layer chromatography and spectrophotometry of a-keto acid hydrazones, Biochim. et biophys. acta 78:85-90, 1963.

9. Shear, C. S., Nyhan, W. L., and Tocci, P. M.: Tyrosinosis and Tyrosinemia in Nyhan, W. L. (ed.): *Amino Acid Metabolism and Genetic Variation* (New York: McGraw-Hill Book Company, Inc., 1967).

10. Bloxam, H. R., Day, M. G., Gibbs, N. K., and Woolf, L. I.: An inborn defect in the metabolism of tyrosine in infants on a normal diet, Biochem. J. 77:320-326, 1960.

11. Lonsdale, D., Faulkner, W. R., Price, J. W., and Smeby, R. R.: Intermittent cerebellar ataxia associated with hyperpyruvic acidemia, hyperalaninemia, and hyper-alaninuria, Pediatrics 43:1025-1034, 1969.

12. Tada, K., Yoshida, T., Konno, T., Wada, Y., Yokoyama, Y., and Arakawa, T.: Hyperalaninemia with pyruvicemia, Tohoku J. Exper. Med. 97:99-100, 1969.

13. Perry, T. L., Hardwick, D. F., Dixon, G. H., Dolman, C. L., and Hansen, S.: Hypermethioninemia. A metabolic disorder associated with cirrhosis, islet cell hyperplasia, and renal tubular degeneration, Pediatrics 36:236-250, 1965.

14. Buist, N. R. M.: Set of simple side-room urine tests for detection of inborn errors of metabolism, Brit. M. J. 2:745-749, 1968.

15. Renuart, A. W.: Screening for inborn errors of metabolism associated with mental deficiency or neurologic disorders or both, New England J. Med. 274:384-387, 1966.

16. Snyderman, S. E.: Diagnosis of metabolic disease, Pediat. Clin. North America 18:199-208, 1971.

17. Neish, W. J. P.: a-Keto Acid Determinations in Glick, D. (ed.): *Methods of Biochemical Analysis* (Vol. 5; New York: Interscience Publishers, Inc., 1957).

18. Datta, S. P., Harris, H., and Rees, K. R.: Chromatography of 2:4 dinitrophenyl-hydrazones of keto acids on alumina, Biochem. J. 46:36, 1950.

19. Cavallini, D., Frontali, N., and Toschi, G.: Keto-acid content of human blood and urine, Nature 164:792-793, 1949.

20. Seligson, D., and Shapiro, B.: Alpha-keto acids in blood and urine studied by paper chromatography, Anal. Chem. 24:754-755, 1952.

21. Denti, E., and Luboz, M. P.: Separation of 2,4-dinitrophenylhydrazones of carbonyl compounds by thin-layer chromatography, J. Chromatogr. 18:325-330, 1965.

22. Henry, R. J.: *Clinical Chemistry: Principles and Technics* (New York: Harper & Row, Publishers, Inc., 1964).

23. Wright, S. W.: Phenylketonuria, J.A.M.A. 165:2079-2083, 1957.

24. Centerwall, W. R., Chinnock, R. F., and Pusavat, A.: Phenylketonuria. Screening programs and testing methods, Am. J. Pub. Health 50:1668-1677, 1960.

25. Woolf, L. I.: Inherited metabolic disorders. Errors of phenylalanine and tyrosine metabolism, Adv. Clin. Chem. 6:97-230, 1963.

26. Dancis, J., and Levitz, M.: Abnormalities of Branched-Chain Amino Acid Metabolism in Stanbury, J. B., Wyngaarden, J. B., and Fredrickson, D. S. (eds.): *The Metabolic Basis of Inherited Disease* (3rd ed.; New York: McGraw-Hill Book Company, Inc., 1972).

CHAPTER 5

Tyrosyluria, Tyrosinemia and Tyrosinosis: Nitrosonaphthol Test

There are several clinical disorders associated with a marked alteration in the metabolism of tyrosine by the liver. As discussed in Chapter 3, these conditions are each associated with a marked accumulation of tyrosine and its metabolites (Table 5-1).

These conditions include neonatal tyrosinemia (1), advanced liver damage secondary to a variety of disorders (2), tyrosinosis (Medes disease) (3), tyrosinemia with hepatorenal disease (4,5), and a more recently recognized disorder of tyrosinemia lacking liver and renal involvement (6,7).

One of the common features of each of these disorders is a marked increase in the urinary concentration of one or more tyrosine metabolites. These metabolites include p-hydroxyphenylpyruvic acid, p-hydroxyphenyllactic acid, and p-hydroxyphenylacetic acid.

Specificity

Perry et al., in a report describing a number of urinary screening tests, gave details on the use of nitrosonaphthol as a means of detecting patients with "tyrosinemia" (8). This reaction is not specific for typosine and it is now known that metabolites of tyrosine, such as p-hydroxyphenylpyruvate, p-hydroxyphenyllactate and p-hydroxyphenylacetic acid also react in this test (8,9). Udenfriend et al. have also shown that 5-hydroxyindoles will react with nitroso-2-naphthol in dilute HCl or H_2SO_4 with traces of nitrite to give a violet chromophore (10,11). Current evidence suggests that the reaction is specific for para-substituted phenols in which the ortho position is either not substituted or has a methyl group (Table 5-2).

TABLE 5-1.–Disorders Associated With an Increased
Urinary Excretion of Tyrosine Metabolites

	Disorder	References
1.	Tyrosinosis–Medes disease	3
2.	Hereditary tyrosinemia associated with hepatorenal disease	4,5
3.	Hereditary tyrosinemia without hepatorenal disease	6,7
4.	Transient neonatal tyrosinemia	1
5.	Severe liver dysfunction	2
6.	Fructosemia	14,15
7.	Galactosemia	16

TABLE 5-2.—Color Reaction Produced With Nitrosonaphthol and Some Frequently Encountered Metabolites in Genetic-Metabolic Disease*

Compound	Reaction
p-Hydroxyphenylpyruvic acid	Red
p-Hydroxyphenyllactic acid	Red
p-Hydroxyphenylacetic acid	Red
p-Hydroxyphenylpropionic acid	Red
Tyrosine	Red
Homogentisic acid	Negative
Phenylpyruvic acid	Negative
Imidazolepyruvic acid	Negative
a-Ketoisocaproic acid	Negative
a-Ketoisovaleric acid	Negative

*100 Mg of each metabolite was dissolved in 100 ml of distilled water.

Modifications

Thomas modified the original procedure by increasing the acidity with HCl, diluting the nitric acid and controlling the temperature so as to overcome the instability of the color which was a serious objection in the original procedure (12).

In yet another attempt to overcome the instability of the red complex, Udenfriend and Cooper heated the solutions of para phenol compounds in the presence of nitric acid and nitrosonaphthol for 30 minutes (13). This procedure resulted in the conversion of the unstable red complex to a stable yellow complex which after the removal of excess nitrosonaphthol could be measured in a photometer.

Udenfriend and Cooper also added small amounts of sodium nitrite to the reaction mixture to overcome the inhibition of the reaction which occurred under some conditions with trichloroacetic acid filtrates (13). Perry et al. used sodium nitrite in the reaction but did not use the heating step (8). Run in this fashion, the presence of p-hydroxy compounds result in an orange red color. Due to the formation of a red color in the procedure of Perry, it is not necessary to remove the unreacted nitrosonaphthol to interpret the results of the test.

Nitrosonaphthol Test

Method of Perry et al. (8)

Reagents:

1. Nitric acid, 2.63N: Add 10 ml of concentrated nitric acid to 50 ml of water.

2. Sodium nitrite, 2.5%: Dissolve 2.5 Gm of sodium nitrite in water and dilute to 100 ml with water.

3. 1-nitroso-2-naphthol,* 0.1%: Dissolve 100 mg of 1-nitroso-2-naphthol in 95% ethanol and dilute to 100 ml with 95% ethanol.

---

*Source: Eastman Organic Chemicals, Rochester 3, New York.

TABLE 5-3.—Sensitivity of the Nitrosonaphthol Test*

1.25 mg/100 ml	Yellow	Negative
6.3 mg/100 ml	Orange	1+
12.5 mg/100 ml	Orange	1+
25.0 mg/100 ml	Orange-red	2+
50.0 mg/100 ml	Orange-red	2+
75.0 mg/100 ml	Red	3+
100.0 mg/100 ml	Deep red	4+

*The test compound was p-hydroxyphenylpropionic acid.

4. Store the sodium nitrite and the nitrosonaphthol solutions in a refrigerator (8). The nitrosonaphthol solution can be stored for at least 2 months in an amber bottle in a refrigerator below 5°C.

Procedure:

1. Place 1 ml of 2.63N nitric acid in a test tube.

2. Add one drop of sodium nitrite.

3. Add 0.1 ml of nitrosonaphthol reagent.

4. Mix the above and then add 0.15 ml of urine.

5. Mix well.

6. Observe for presence of color for 5 minutes.

Results:

1. A positive test is indicated by the formation of an orange-red color within 2-5 minutes.

2. A negative test is indicated by the persistence of the original yellow color for at least 5 minutes. The development of a red color after 5 minutes is not taken as a positive result.

3. The results of this test are reported as 1+ to 4+ as shown in Table 5-3.

References

1. Kretchmer, N., Levine, S. Z., and McNamara, H.: The in vitro metabolism of tyrosine and its intermediates in the liver of the premature infant, Am. J. Dis. Child. 93: 19, 1957.

2. Felix, K., Leonhardi, G., and Glasenapp, I. V.: Uber Tyrosinosis, Hoppe-Seyler's Zeitschrift für Physiologische Chemie. 287:141-147, 1951.

3. Medes, G.: A new error of tyrosine metabolism: tyrosinosis. The intermediary metabolism of tyrosine and phenylalanine, Biochem. J. 26:917-940, 1932.

4. Gentz, J., Jagenburg, R., and Zetterström, R.: Tyrosinemia. An inborn error of tyrosine metabolism with cirrhosis of the liver and multiple renal tubular defects. (de-Toni-Debré-Fanconi syndrome), J. Pediat. 66:670-696, 1965.

5. Halvorsen, S., Pande, H., Løken, A. C., and Gjessing, L. R.: Tyrosinosis. A study of 6 cases, Arch. Dis. Childhood 41:238-249, 1966.

6. Wadman, S. K., Van Sprang, F. J., Maas, J. W., and Ketting, D.: An exceptional case of tyrosinosis, J. Ment. Def. Res. 12:269-281, 1968.

7. Holston, J. L., Levy, H. L., Tomlin, G. A., Atkins, R. J., Patton, T. H., and Hosty, T. S.: Tyrosinosis. A patient without liver or renal disease, Pediatrics 48: 393-400, 1971.

8. Perry, T. L., Hansen, S., and MacDougall, L.: Urinary screening tests in the prevention of mental deficiency, Canad. M. A. J. 95:89-95, 1966.

9. McCormick, B., Young, S. K., and Woods, M. N.: Specificity of the colorimetric assay of tyrosine with 1-nitroso-2-naphthol, Clin. chim. acta 12:216-218, 1965.

10. Udenfriend, S., Weissbach, H., and Clark, C.: The estimation of 5-hydroxytryptamine (serotonin) in biological tissues, J. Biol. Chem. 215:337-344, 1955.

11. Udenfriend, S., Titus, E., and Weissbach, H.: The identification of 5-hydroxy-3-indoleacetic acid in normal urine and a method for its assay, J. Biol. Chem. 216:499-505, 1965.

12. Thomas, L. E.: A new method for the determination of tyrosine and its use in determining the tyrosine content of edestin, casein, and tobacco mosaic virus, Arch. Biochem. 5:175-180, 1944.

13. Udenfriend, S., and Cooper, J. R.: The chemical estimation of tyrosine and tyramine, J. Biol. Chem. 196:227-233, 1952.

14. Lindeman, R., Gjessing, L. R., Merton, B., and Halvorsen, S.: Fructosaemia "Acute-tyrosinosis," Lancet 1:891, 1969.

15. Lindemann, R., Gjessing, L. R., Merton, B., Löken, A. C., and Halvorsen, S.: Amino acid metabolism in hereditary fructosemia, Acta paediat. scandinav. 59:141-147, 1970.

16. Howell, R. R., and Thomas, G. H.: Unpublished data.

CHAPTER 6

Cystinurias, Homocystinuria and Mercaptolactate Disulfiduria: The Nitroprusside Test

The nitroprusside test has proved to be a useful screening test for at least three genetic metabolic diseases of clinical importance (Table 6-1). Those disorders which are now known to yield a positive result with this test are the cystinurias, homocystinuria and a recently described abnormality characterized by a marked increase in the excretion of β-mercaptolactate-cysteine disulfide (1-3).

Cystinurias

The nitroprusside screening test was originally employed to detect cystinuria, a group of inherited defects of amino acid transport. Cystinuria is now known to be associated with not only an impaired renal tubular reabsorption of cystine, but also of lysine, arginine and ornithine. As a result of this defect these four amino acids are excreted in abnormally large amounts throughout the life of affected homozygous patients. The least soluble of these amino acids, cystine, forms urinary calculi.

It has generally been believed that the clinical consequences of this disorder are limited to those associated directly with the urinary cystine calculi. More recent findings, however, have led to the suggestion that this disorder might be associated with an increased incidence of mental retardation (4). Additional data are required, however, to determine the validity of this suggestion.

Harris and co-workers have presented evidence that patients with cystinuria can be divided into at least two groups: one in which the disorder presents as a true autosomal recessive and a second in which it presents as an incompletely recessive disorder (5). The second group differs from the first group in that the heterozygous carrier often excretes increased amounts of both

TABLE 6-1.—Inborn Errors of Metabolism Associated With Positive
Nitroprusside Reactions

	Disorder	Major Reactive Compounds	Color Reaction
1.	Cystinuria: type 1 homozygous patient	Cystine	Purple
2.	Cystinuria: types 2 & 3 homozygous patient & heterozygous carrier	Cystine	Purple
3.	Homocystinuria	Homocystine	Purple
4.	β-Mercaptolactate-cysteine disulfiduria	β-Mercaptolactate-cysteine-disulfide	Purple

cystine and lysine thus yielding a positive nitroprusside reaction. Additional evidence has indicated that this second group of patients, may in fact, be further divided into at least two distinct genetic disorders (6).

Homocystinuria

The second well-documented disorder which yields a positive result with the nitroprusside test is homocystinuria. The positive reaction in this genetic abnormality results, however, not from cystine but rather, as the name suggests, from an increased urinary concentration of homocystine (2).

In the classical and most studied form of this disease the increase in the urinary concentration of homocystine results from a deficiency or absence of cystathionine synthase activity (7). The major clinical consequences of this autosomal recessive disorder are osteoporosis, ectopia lentis, vascular disease and a high frequency of mental retardation.

More recently a number of additional genetic defects associated with increased levels of urinary homocystine have been recognized. These include at least two distinct abnormalities affecting the utilization of vitamin B_{12} required for the normal methylation of homocystine (8-11) as well as another form of cystinuria associated with decreased methylenetrahydrofolate reductase activity (12).

β-Mercaptolactate-Cysteine Disulfiduria

The third and most recently described disorder yielding a positive result with this test has been that found in a single patient by Ampola et al. (3). This patient was a 47-year-old, severely retarded male who was found to have a positive nitroprusside test in urine in the absence of either cystinuria or homocystinuria. The positive reaction was subsequently found to be the result of an increased excretion of β-mercaptolactate-cysteine disulfide (13). As only one patient has been reported with this disorder its importance as a cause of mental retardation is not known. The fact, however, that additional cases have not been detected in large scale screening programs of retarded patients, combined with the observation that this patient was the product of a sibling mating, indicates that it is very rare.

Chemical Bases of Reaction

As can be seen from the above discussion, each of these disorders is quite dissimilar, both in terms of the basic biochemical defect and in clinical consequences. They are related only in that the urine of patients with each of these disorders will yield a positive cyanide-nitroprusside reaction. This results from the fact that each of these conditions is specifically associated with a marked increase in the urinary concentration of compounds that are capable of yielding free sulfhydryl groups upon chemical reduction by a solution of sodium cyanide.

It should be noted that while each of these disorders is associated with a marked increase in the excretion of the oxidized disulfides, the sodium nitroprusside reacts with only the reduced free sulfhydryl group. The screening test described here utilizes a 5% solution of sodium cyanide to bring about the reduction of each of these compounds. The final color development upon the addition of the sodium nitroprusside is thus the net result of both the available reducible disulfides plus any preformed free sulfhydryl groups already present in the urine. As a result, the cyanide-nitroprusside reaction is not a specific reaction, but is rather a general test for any compound containing a sulfhydryl group or being capable upon reduction by cyanide of yielding this group.

Specificity

In addition, sodium nitroprusside has also been used as a test for the presence of creatinine, acetone and acetoacetic acid (14,15). The color formation in the presence of these compounds is reported, however, to occur only in the presence of a strong solution of alkali, i.e., sodium hydroxide. The use of NH_4OH instead of $NaOH$ is reported to prevent interference from creatinine (16). In connection with this observation, Brand et al. used a modification of the nitroprusside test in which alkali was omitted altogether for the stated reason that both creatinine and/or acetone would give only a "very faint" reaction in the absence of alkali, while the sulfhydryl groups would still give a very strong reaction (17).

The test employed here, however, employs NH_4OH and thus, in theory at least, acetone and acetoacetic acid might still remain a potential source of confusion. Our own experience, however, indicates that this is not the case. We find that acetoacetic acid and acetone solutions yield negative results with this particular version of the nitroprusside test.

Silver-Nitroprusside Modification

A very useful modification of the urinary nitroprusside test has been introduced by Spaeth and Barber (18). This modification allows one to differentiate between cystine and homocystine. This is accomplished by substituting silver nitrate for the sodium cyanide as the reducing agent. The resulting specificity results from the fact that homocystine is reduced to the reactive form (homocysteine) by silver nitrate, while cystine remains in the nonreactive oxidized form. The homocysteine then reacts with the sodium nitroprusside to yield a positive result. This test is reported to be not only specific for homocystine (and homocysteine), but also very sensitive since it can yield a positive result with as little as 2 mg/100 ml of homocysteine. We have found the differentiation of homocystine from cystine quite good by this test.

Nitroprusside Test

Procedure Given by Knox (19)

Reagents:
1. Ammonium hydroxide, concentrated.
2. Sodium cyanide, 5%: Add 5 Gm. of NaCN to distilled water and dilute to a total volume of 100 ml.
3. Sodium nitroprusside (sodium nitroferricyanide) 5%: Add 5 Gm of sodium nitroprusside to distilled water and dilute to 100 ml total volume.

Sample:
1. Urine, 5 ml.

Procedure:
1. Add 5 drops of concentrated ammonium hydroxide to 5 ml of urine.
2. Add 2 ml of freshly prepared 5% NaCN and mix well.
3. Allow to stand for 10 minutes.
4. Add 5 drops of freshly prepared sodium nitroprusside, mix well, and observe for color.

TABLE 6-2.—The Sensitivity of the Nitroprusside Test With Cystine

Cystine Concentration	Color	Result
3.125 mg/100 ml	Yellow	Negative
6.25 mg/100 ml	Yellow-pink	Trace
12.5 mg/100 ml	Pink	+
25 mg/100 ml	Purplish-pink	++
50 mg/100 ml	Purple	+++
100 mg/100 ml	Dark purple	++++

Results:

1. A positive result for cystine or homocystine is a deep purple color which fades gradually (20).

2. Normal urines give a pale brown and occasionally a very faint flesh color.

3. It has been found that the concentration of homocystine in homocystinuria patients is sometimes low and it has been suggested that with low protein intake and large urine volumes, the nitroprusside test might fail to detect homocystinuria (20).

4. The sensitivity of this test using cystine as the standard is shown in Table 6-2.

5. Do not add NaCN to an acid solution.

Silver Nitroprusside Test

Procedure of Spaeth, G., and Barber, G. (18)

Reagents:

1. Sodium chloride.

2. Ammonia, 3%: Dilute 5 ml of concentrated ammonium hydroxide (28%-30%) to a total of 50 ml with distilled water.

3. Sodium nitroprusside (sodium nitroferricyanide), 1%. Add 100 mg of sodium nitroprusside to water and dilute to a final volume of 10 ml.

4. Silver nitrate ($AgNO_3$), 1.0% in 3% ammonia: Add 200 mg of silver nitrate to a solution of 3% ammonium and dilute to a final volume of 20 ml with 3% ammonia.

5. Sodium cyanide, 0.7%. Dilute 70 mg of sodium cyanide to a final volume of 10 ml of distilled water.

Method:

1. Place 6-7 ml of urine in a test tube.

2. Saturate the above urine sample with solid chloride.

3. The above sample is treated in the following manner:

Reagent	Control	Test
NaCl-Saturated urine	2.5 ml	2.5 ml
3% Ammonia (no silver)	0.25 ml	- - -
1.0% Silver nitrate in 3% ammonia	- - -	0.25 ml

4. Allow to stand 1 minute and then add the following:

1.0% Sodium nitroprusside	0.25 ml	0.25 ml
0.7% Sodium cyanide	0.25 ml	0.25 ml

5. Observe for the *immediate* development of pink or purple color.

Results:

1. The *immediate* formation of a pink or purple color is taken as a positive result.

2. This test is positive for homocystine while a negative result is obtained with cystine.

Notes:

1. Cystine will undergo a slow reaction resulting in a positive test after a sufficient period of time. This is not to be confused with the immediate reaction given by homocystine.

2. Cyanide addition is required to bind the silver ion and thus allow for the formation of the colored product by the reaction between homocysteine and nitroprusside.

3. This modification of the urinary nitroprusside test was introduced to differentiate between cystine and homocystine.

References

1. Lewis, H. B.: The occurrence of cystinuria in healthy young men and women, Ann. Int. Med. 6:183-192, 1932-33.

2. Carson, N. A. J., Cusworth, D. C., Dent, C. E., Field, C. M. B., Neill, D. W., and Westall, R. G.: Homocystinuria. A new inborn error of metabolism associated with mental deficiency, Arch. Dis. Childhood 38:425-436, 1963.

3. Ampola, M. G., Efron, M. L., Bixby, E. M., and Meshorer, E.: Mental deficiency and a new aminoaciduria, Am. J. Dis. Child. 117:66-70, 1969.

4. Scriver, C. R., Whelan, D. T., Clow, C. L., and Dallaire, L.: Cystinuria. Increased prevalence in patients with mental disease, New England J. Med. 283:783-786, 1970.

5. Harris, H., Mittwoch, U., Robson, E. B., and Warren, F. L.: Phenotypes and genotypes in cystinuria, Ann. Human Genetics 20:57-91, 1955-56.

6. Thier, S. O., and Segal, S.: Cystinuria in Stanbury, J. B., Wyngaarden, J. B., and Fredrickson, D. S. (eds.): *The Metabolic Basis of Inherited Disease* (3rd ed.; New York: McGraw-Hill Book Company, 1972.)

7. Mudd, S. H., Finkelstein, J. D., Irreverre, F., Laster, L.: Homocystinuria. An enzymatic defect, Science 143: 1443-1445, 1964.

8. Mudd, S. H., Levy, H. L., and Morrow, II, G.: Deranged B_{12} metabolism. Effects on sulfur amino acid metabolism, Biochem. Med. 4:193-214, 1970.

9. Mudd, S. H., Uhlendorf, B. W., Hinds, K. R., and Levy, H. L.: Deranged B_{12} metabolism. Studies of fibroblasts grown in tissue culture, Biochem. Med. 4:215-239, 1970.

10. Goodman, S. I., Moe, P. G., Hammond, K. B., Mudd, S. H., and Uhlendorf, B. W.: Homocystinuria with methylmalonic aciduria. Two cases in a sibship, Biochem. Med. 4:500-515, 1970.

11. Hollowell, J. G., Jr., Hall, W. K., Coryell, M. E., McPherson, J., Jr., and Hahn, D. A.: Homocystinuria and organic aciduria in a patient with vitamin B_{12} deficiency, Lancet 2:1428, 1969.

12. Mudd, S. H., Uhlendorf, B. W., Freeman, J. M., Finkelstein, J. D., and Shih, V. E.: Homocystinuria associated with decreased methylenetetrahydrofolate, reductase activity, Biochem. Biophys. Res. Comm. 46:905-912, 1972.

13. Crawhall, J. C., Parker, R., Sneddon, W., and Young, E. P.: β-mercaptolactate-cysteine disulfide in the urine of a mentally retarded patient, Am. J. Dis. Child. 117:71-82, 1969.

14. Rothera, A. C. H.: Note on the sodium nitro-prusside reaction for acetone, J. Physiol. 37:491-494, 1908.

15. Legal, E.: Über eine neueAceton Reaction und deren Verwendbarkeit zur Harnuntersuchung, Breslau. Arztl. Z. 5:25-27 and 38-40, 1823.

16. Sullivan, M. X.: A distinctive test for cysteine, Pub. Health Rep. 41:1030-1056, 1926.

17. Brand, E., Harris, M. M., and Biloon, S.: The excretion
of a cystine complex which decomposes in the urine
with the liberation of free cystine, J. Biol. Chem. 86:
315-331, 1930.

18. Spaeth, G. L., and Barber, G. W.: Prevalence of homocystinuria among the mentally retarded: Evaluation of
a specific screening test, Pediatrics 40:586-589, 1967.

19. Knox, W. E.: Cystinuria in Stanbury, J. B., Wyngaarden, J. B., and Fredrickson, D. S. (eds.): *The Metabolic Basis of Inherited Disease* (2nd ed.; New York:
McGraw-Hill Book Company, Inc., 1960).

20. Schimke, R. N., McKusick, V. A., Huang, T., and Pollack, A. D.: Homocystinuria. Studies of 20 families
with 38 affected members, J.A.M.A. 193:711-719, 1965.

CHAPTER 7

Mucopolysaccharidoses: Turbidity and Spot Tests

The acid mucopolysaccharides are widely distributed in various tissues and fluids of the body. This class of compounds includes hyaluronic acid, chondroitin sulfuric acid A, chondroitin sulfuric acid B, chondroitin sulfuric acid C, chondroitin, keratosulfate, heparin and heparin sulfuric acid. The detection of these compounds has become increasingly important with the findings that there are several inherited disorders associated with increased concentrations of the acid mucopolysaccharides in urine (1,2). McKusick has presented evidence which indicates that this group of disorders can be divided into at least eight distinct entities (Table 7-1). The clinical aspects of these disorders have been reviewed in detail elsewhere (2,3,4).

TABLE 7-1.—The Genetic Mucopolysaccharidoses (As Classified in 1972)*

Designation		Clinical Features	Genetics	Excessive Urinary MPS	Substance Deficient
MPS I H	Hurler syndrome	Early clouding of cornea, grave manifestations, death usually before age 10	Homozygous for MPS IH gene	Dermatan sulfate Heparan sulfate	a-L-Iduronidase (formerly called Hurler corrective factor)
MPS I S	Scheie syndrome	Stiff joints, cloudy cornea, aortic regurgitation, normal intelligence, ?normal life span	Homozygosity for MPS IS gene	Dermatan sulfate Heparan sulfate	a-L-Iduronidase
MPS I H/S	Hurler-Scheie compound	Phenotype intermediate between Hurler and Scheie	Genetic compound of MPS I H and I S genes	Dermatan sulfate Heparan sulfate	a-L-Iduronidase
MPS II A	Hunter syndrome, severe	No clouding of cornea, milder course than in MPS IH but death usually before age 15 years	Hemizygous for X-linked gene	Dermatan sulfate Heparan sulfate	Hunter corrective factor
MPS II B	Hunter syndrome, mild	Survival to 30s to 50s, fair intelligence	Hemizygous for X-linked allele for mild form	Dermatan sulfate Heparan sulfate	Hunter corrective factor
MPS III A	Sanfilippo syndrome A	Identical phenotype	Homozygous for Sanfilippo A gene	Heparan sulfate	Heparan sulfate sulfatase
MPS III B	Sanfilippo syndrome B	Mild somatic, severe central nervous system effects	Homozygous for Sanfilippo B gene (at different locus)	Heparan sulfate	N-Acetyl 1-a-D-galactosaminidase
MPS IV	Morquio syndrome	Severe bone changes of distinctive type, cloudy cornea, aortic regurgitation	Homozygous for Morquio gene	Keratan sulfate	Unknown
MPS V	Vacant				

*From McKusick, V.: *Heritable Disorders of Connective Tissue* (4th ed.; St. Louis: C. V. Mosby Company), in press.

Screening Tests

With the recognition of these disorders, several screening procedures have been developed to detect the presence of abnormal amounts of the mucopolysaccharides in random urine samples. The majority of these tests are based on one of the following reactions:

1. Reaction of the urinary mucopolysaccharides with acid albumin to form a turbid solution and/or precipitate.
2. Reaction of the urinary mucopolysaccharides with quarternary ammonium salts to form a turbid solution and/or precipitate.
3. Reaction of a variety of basic dyes with the mucopolysaccharides in urine spotted on filter paper.

Acid Albumin Turbidity Test

One of the more useful methods is the so-called acid albumin turbidity test. This test was introduced by Meyer and Palmer after they noted that a "vitreous humor carbohydrate" preparation formed a precipitate with serum globulins in the presence of acid (5). In 1939, C. V. Seastone used this test as a basis for the quantitative estimation of mucopolysaccharides derived from group C streptococci (6). This procedure was also utilized to measure hyaluronidase activity against hyaluronic acid as the turbidity, which is formed in the presence of acid albumin, is proportional to the hyaluronic acid concentration (7,8).

Dorfman and Ott later studied the relationship of the optical density to hyaluronic acid concentration, albumin concentration, pH, ionic strength and time of reading (9). It was later shown that, with only minor modifications, this test could be used for the estimation of the acid mucopolysaccharides in urine (10).

A number of modifications of this procedure have since been employed as screening tests for patients with the various mucopolysaccharidoses. One of the simplest is that of Denny and Dutton in which four drops of a 20% bovine albumin solution is added to an inch of urine in a small test tube (11). This solution is mixed and a 10% acetic acid solution is added drop by drop. A positive test is indicated by a thick white precipitate. In a similar procedure, given by McKusick, 5 ml of an acidified bovine serum albumin (0.1% bovine serum albumin in 0.1M acetate buffer pH 3.75) is added to 1 ml of urine (12). In this test, as before, a positive result is indicated by the development of a dense white turbidity.

As the turbidity reaction is extremely sensitive to variations in the ionic strength, several methods employ dialyzed urine for the test (10,13). This, combined with the photometric estimation of the turbidity, results in an increased standardization and reproducibility of the results.

Cetyltrimethyl-Ammonium Bromide

Several procedures utilizing cetyltrimethyl-ammonium bromide, in place of the acid albumin, have also been used for the detection of abnormal amounts of the acid mucopolysaccharides (14,15). These procedures are based on the fact that cetyltrimethyl-ammonium bromide yields water-insoluble quarternary salts with the sulphated and nonsulphated acidic mucopolysaccharides (17). As with the acid albumin procedures, these methods also result in the formation of turbidity and/or precipitate in the presence of excessive amounts of the mucopolysaccharides. A similar method has also been utilized by Manley and Hawksworth (16). In this procedure, however, the test is based on the formation of an insoluble complex between the mucopolysaccharides and cetylpyridinium chloride.

Metachromatic Spot Tests

In addition to the various screening procedures based on the production of turbidity, several tests have also been developed which take advantage of the metachromatic staining properties of the mucopolysaccharides. Although toluidine blue had been utilized as a means to locate chondroitin sulfuric acid on chromatographs of urine (18), Berry and Spinanger concluded that since urines from normal controls gave a negative result when tested directly with the stain that the chromatographic separation was unnecessary for screening purposes (19). With this in mind, a direct filter paper test was devised to detect patients with increased mucopolysaccharide excretion (19).

With this procedure small amounts of urine are spotted on filter paper and then stained with toluidine blue. The staining of the paper is performed at a pH of 2.0 to prevent the reaction of compounds which are ionized above this pH, thus making the procedure more specific. It should be noted, however, that many other substances beside the mucopolysaccharides can produce metachromasia, even at this pH.

A number of similar procedures have also been developed and found useful as screening tools (12,20-22). One of these procedures, devised by Dr. Elaine Berman, employs a paper previously impregnated with Azure A on which the urine is then spotted, dried and destained. These papers are manufactured by Ames-Yissum, Ltd., Jerusalem, Israel, and are available in the U.S.A. from the Ames Company, Elkhart, Indiana. We found results with these papers to be similar to those obtained with the toluidine spot test. They would thus appear to be useful for the occasional user, but expensive when used with a large number of urines.

Results

A comparison of the results obtained by the various screening tests has been made by several groups (23-26). As each of these investigators arrived at a somewhat different conclusion, it is difficult to recommend any one test.

The results do, however, indicate that each of these tests yield on occasion, "false negative" and/or "false positive results." It is, therefore, strongly recommended that more than one urine sample should be tested before any conclusion is made regarding the results of the test. This is particularly true when borderline results are obtained. We have also found it to be useful to utilize more than one procedure, especially when questionable results are obtained with a particular test. We have also found that accurate results are obtained only when one uses very *fresh urine* samples. The incidence of "false negative results" increases markedly with the age of the urine sample (when not frozen) presumably secondary to bacterial contamination.

We list three tests which, in our hands, separate patients with the mucopolysaccharidoses from normals in virtually every situation. We have studied all known types of mucopolysaccharidoses with these tests. It should also be stressed, however, that these procedures should be utilized as screening tests only and that the final diagnosis should rest upon other procedures (27).

Toluidine Blue Test for Mucopolysaccharides

Modification of the Method Given by McKusick (12)

Reagent:

1. Toluidine blue reagent: Dissolve 1.0 Gm of toluidine blue in a solution of 400 ml of acetone and 100 ml of water.

2. Acetic acid, 10%.

Specimen:

1. Urine: 1 ml, very fresh.

Procedure:

1. Add 10 and 25 μl of urine to spots on a small piece of filter paper using 5 μl increments, and allow each to dry before adding the next sample.
2. Dip papers into the toluidine blue reagent for 45 seconds and then drain well.
3. Wash in two separate portions of 10% acetic acid.

Result:

1. A purple spot against a blue background is seen in a positive test. A faint rim of metachromasia around the dried spot is considered negative.
2. Normal newborn infants might, on occasion, give a positive result.

Notes:

1. The method given by McKusick (12) states that 10 and 25 ml amounts of urine should be placed on filter paper, however, this should have read as 10 and 25 μl amounts of urine.
2. We use this modification rather than the original Berry method.

Cetyltrimethylammonium Bromide Turbidity Test

Method of Renuart, A. W. (14)

Reagents:

1. Citrate buffer, 1M, pH 6.0: Dissolve 105 Gm of citric acid monohydrate in 250 ml of water. Slowly add 75.0 ml of 19N sodium hydroxide. Mix well and allow to cool to room temperature. Measure pH with a pH meter and adjust to a final pH of 6.0 with the addition of sodium hydroxide. Dilute to a final volume of 500 ml (15).
2. Cetyltrimethylammonium bromide reagent, 5%: Dissolve 5 Gm of cetyltrimethylammonium bromide (hexadecyltrimethylammonium bromide*) in the 1M citrate buffer, pH 6.0, and dilute to a final volume of 100 ml with the same buffer.

Specimen:

1. Urine: 10 ml (very fresh).

Procedure:

1. Place 5 ml of urine in a test tube.
2. Allow the urine to stand until it is at room temperature (Important!).
3. Add 1 ml of cetyltrimethylammonium reagent.
4. Observe for turbidity at end of 30 minutes.

---

*Source: Eastman Organic Chemicals, Rochester 3, New York.

Results:

1. A cloudy precipitate constitutes a positive test.

Notes:

1. Urine must be at room temperature as it has been reported that "a cold urine will invariably give a positive test" (14).

2. Repeated studies on patients with "gargoylism" were always found to be positive with this test; however, in the same study there was also a 1% false positive rate in a study of 2,000 patients with mental deficiency or other neurologic disorders (14).

3. This test has been found to give a heavier precipitate with chondroitin sulfate B than with an equal concentration of heparitin sulfate.

Acid Albumin Turbidity

Method of Dorfman (2,10)

Reagents:

1. Phosphate-citrate buffer, 0.3M with 0.45M NaCl, pH 5.6.

   a. Citric acid, 0.6M: Dissolve 11.5 Gm grams of citric acid in water and dilute to a final volume of 100 ml.

   b. Sodium phosphate, dibasic, 0.6M: Dissolve 8.5 Gm of anhydrous Na_2HPO_4 (or 16.1 Gm of $Na_2HPO_4 \cdot 7H_2O$ or 21.5 Gm of $Na_2HPO_4 \cdot 12H_2O$) in distilled water; heat to dissolve and dilute to a final volume of 100 ml.

   c. Add 21 ml of the 0.6M citric acid to 29 ml of the sodium phosphate. Adjust pH to 5.6.

   d. Dissolve 2.68 Gm of NaCl in the solution prepared above (part c) and dilute to a final volume of 100 ml with distilled water.

2. Acetate buffer, 0.1M, pH 3.75.

   a. Acetic acid, 0.2M: Place 1.15 ml of glacial acetic acid in distilled water and dilute to 100 ml with distilled water.

   b. Sodium acetate, 0.2M: Dissolve 2.72 Gm of sodium acetate ($NaC_2H_3O_2 \cdot 3H_2O$) or 1.64 Gm of sodium acetate ($NaC_2H_3O_2$) in water and dilute to 100 ml with distilled water.

   c. Add 46.3 ml of the acetic acid (part a) to 3.7 ml of the sodium acetate (part b). If required adjust to pH 3.75. Dilute to a final volume of 100 ml with distilled water.

3. Acid albumin reagent: Dissolve 50 mg of bovine serum albumin in the 0.1M acetate buffer pH 3.75 and dilute to 50 ml volume with the same buffer. This solution can be stored at 4°C.

Specimen:

1. Urine: 5 ml (very fresh).

Procedure:

1. Centrifuge urine for 5 minutes in a clinical centrifuge.

2. Remove the supernatant and dialyze against distilled water at 5°C for at least 3 hours. Change water at least one time during this period of time.

3. Allow urine and reagents to come to room temperature.

4. Place 0.75 ml of the dialyzed urine in a clean test tube. Add 0.25 ml of 0.3M phosphate-citrate buffer pH 5.6 containing 0.45M NaCl.

5. Add 5 ml of the acid albumin reagent.

6. Prepare a "control" in the same way as above, except use 0.1M acetate buffer pH 3.75 in place of the acid albumin reagent.

7. Determine optical dentity (O.D.) at exactly 10 minutes of both of the above samples at 540 nm using water as a blank.

Results:

1. Subtract the O.D. of control from the O.D. of the sample. This gives the final O.D.

2. Dorfman (2) reports that with a light path of 1 cm optical densities above 0.050 have been found only in Hurler's syndrome or related syndromes. Similar results were found by Steiness (13). Our experiences show similar results. Since this measurement results from reflecting turbidity (light scattering) the data will not however be directly comparable from instrument to instrument.

Notes:

1. It has been found using hyaluronic acid and acidified horse serum albumin, that maximum turbidity develops at pH 3.81. The turbidity is decreased if the pH is allowed to go much above or below this value (9).

2. Turbidity development decreases as ionic strength increases (9). This observation shows the importance of using only dialyzed urine.

3. It appears that the time of reading should be uniform as it has been found that the turbidity increases with time (9) while the difference in optical density between different concentrations decreases (9).

4. This test might yield a negative or weak positive result with urine from patients with the Morquio syndrome, as keratosulfate has no hexuronic acid. This lack of a strong acid group results in a reduced reaction with the various basic salts, i.e., cetyltrimethylammonium bromide, and basic dyes, i.e., toluidine blue.

References

1. Meyer, K., Grumbach, M. M., Linker, A., and Hoffman, P.: Excretion of sulfated mucopolysaccharides in gargoylism (Hurler's syndrome), Proc. Soc. Exper. Biol. & Med. 97:275-279, 1958.

2. Dorfman, A.: Heritable Diseases of Connective Tissues: The Hurler Syndrome in Stanbury, J. B., Wyngaarden, J. B., and Fredrickson, D. S. (eds.): *The Metabolic Basis of Inherited Disease* (2nd ed.; New York: McGraw-Hill Book Company, Inc., 1966.

3. McKusick, V. A., Kaplan, D., Wise, D., Hanley, W. B., Suddarth, S. B., Sevick, M. E., and Maumenee, A. E.: The genetic mucopolysaccharidoses, Medicine 44:445-483, 1965.

4. McKusick, V.: *Heritable Disorders of Connective Tissue* (3rd ed.; St. Louis: C. V. Mosby Company, 1966).

5. Meyer, K., and Palmer, J. W.: On glycoproteins. II. The polysaccharides of vitreous humor and of umbilical cord, J. Biol. Chem. 114:689-703, 1936.

6. Seastone, C. V.: The virulence of Group C hemolytic streptococci of animal origin, J. Exper. Med. 70:361-377, 1939.

7. Kass, E. H., and Seastone, C. V.: The role of the mucoid polysaccharide (hyaluronic acid) in the virulence of group A hemolytic streptococci, J. Exper. Med. 79:319-330, 1944.

8. Leonard, S. L., Perlman, P. L., and Kurzrok, R.: A turbidimetric method for determining hyaluronidase in semen and tissue extracts, Endocrinology 39:261-269, 1946.

9. Dorfman, A., and Ott, M. L.: A turbidimetric method for the assay of hyaluronidase, J. Biol. Chem. 172:367-375, 1948.

10. Dorfman, A.: Studies on the biochemistry of connective tissue, Pediatrics 22:576-589, 1958.

11. Denny, W., and Dutton, G.: Simple urine test for gargoylism, Brit. M. J. 1:1555-1556, 1962.

12. McKusick, V.: *Heritable Disorders of Connective Tissue* (2nd ed.; St. Louis: C. V. Mosby Company, 1960).

13. Steiness, I.: Acid mucopolysaccharides in urine in gargoylism, Pediatrics 27:112-117, 1961.

14. Renuart, A. W.: Screening for inborn errors of metabolism associated with mental deficiency or neurologic disorders or both, New England J. Med. 274:384-387, 1966.

15. Perry, T. L., Hansen, S., and MacDougall, L.: Urinary screening tests in the prevention of mental deficiency, Canad. M. A. J. 95:89-95, 1966.

16. Manley, G., and Hawksworth, J.: Diagnosis of Hurler's syndrome in the hospital laboratory and the determination of its genetic type, Arch. Dis. Childhood 41:91-96, 1966.

tion of its genetic type, Arch. Dis. Childhood 41:91-96, 1966.

17. Bera, B. C., Foster, A. B., and Stacey, M.: Observations on the properties of cetyltrimethylammonium salts of some acidic polysaccharides, J. Chem. Soc. 4:3788-3793, 1955.

18. Lipmann, F.: Biological sulfate activation and transfer, Science 128:575-580, 1958.

19. Berry, H. K., and Spinanger, J.: A paper spot test useful in study of Hurler's syndrome, J. Lab. & Clin. Med. 55:136-138, 1960.

20. Boggs, D. E.: Detection of inborn errors of metabolism, CRC Critical Reviews in Clinical Laboratory Sciences 2:529-572, 1971.

21. Carson, N. A. J., and Neill, D. W.: Metabolic abnormalities detected in a survey of mentally backward individuals in Northern Ireland, Arch. Dis. Childhood 37:505-513, 1962.

22. Tocci, P. M.: The Biochemical Diagnosis of Metabolic Disorders by Urinalysis and Paper Chromatography in Nyhan, W. L. (ed.): *Amino Acid Metabolism and Genetic Variation* (New York: McGraw-Hill Book Company, Inc., 1967).

23. Carter, C. H., Wan, A. T., and Carpenter, D. G.: Commonly used tests in the detection of Hurler's syndrome, J. Pediat. 73:217-221, 1968.

24. Procopis, P. G., Turner, B., Ruxton, J. T., and Brown, D. A.: Screening tests for mucopolysaccharidosis, J. Ment. Defic. Res. 12:13-17, 1968.

25. Pennock, C. A.; Mott, M. G.; and Batstone, G. F.: Screening for mucopolysaccharidoses, Clin. chim. acta 27:93-97, 1970.

26. McDonald, T. P., Lozzio, C., and Lotkin, P.: Rapid detection and identification of mucopolysaccharides in urine, Clinical pediat. 9:272-276, 1970.

27. Frantantoni, J. C., Hall, C. W., and Neufeld, E. F.: The defect in Hurler's and Hunter's syndromes. Faulty degradation of mucopolysaccharide, Proc. Nat. Acad. Sc. 60:699-706, 1968.

CHAPTER 8

Melituria: Screening Tests for Reducing Substances in Urine

Melituria, the increased urinary excretion of any of a variety of simple sugars, can result from several inherited abnormalities in man as shown in Table 8-1. The standard Benedict's test which may be used for the detection of these disorders is based on the ability of excessive amounts of these sugars to reduce copper upon heating for short periods of time (1).

Reducing substances such as glucose, xylulose, galactose, fructose, lactose, glucuronic acid and homogentisic acid will yield a positive reaction with this test. As this screening method does not differentiate between these various compounds a positive result can indicate anyone of several inherited disorders. Clinically these may range in severity from the benign course of pentosuria to the life threatening defects of galactosemia.

Diabetes

Current knowledge does not justify the inclusion of diabetes in a list of inborn errors due to a single gene defect. It is, nevertheless, important that the possibility of this disease be considered whenever one is performing tests for the presence of reducing substances, as the urinary glucose found in this disorder will yield a positive result. Moreover, of course, the presence of

TABLE 8-1.—Disorders of Carbohydrate Metabolism or Transport Which May Be Associated With a Positive Result for Reducing Substances in Urine

Disorder	Compound	References
Diabetes	Glucose	27
Renal glycosuria	Glucose	28
Fanconi syndrome	Glucose	32,33
Glucose-galactose malabsorption	Glucose (intermittent)	29
Pentosuria	Xylulose	2,4
Transferase deficiency galactosemia	Galactose	7,5
Galactokinase deficiency galactosemia	Galactose	8,9
Essential fructosuria	Fructose	15
Fructose intolerance	Fructose	15,16
Lactose intolerance	Lactose	30
Lactase deficiency	Lactose	31

large amounts of glucose in the urine is an abnormal finding and is, therefore, clinically important (see Table 8-1). In our experience the majority of urine samples yielding a positive Benedict's test result for reducing substances are ultimately found to contain excessive amounts of glucose.

Pentosuria

Pentosuria, one of Garrod's original four "inborn errors of metabolism" (2) is now known to be the result of a defect in the conversion of L-xylulose to xylitol by xylitol dehydrogenase (3). This enzymatic defect results in an increased excretion of L-xylulose in the urine of affected individuals. It is the reducing property of this compound which is responsible for the positive Benedict's test result in urine from individuals with this disorder.

From a clinical standpoint the only difficulty one encounters with this biochemical defect is that of the possible confusion with one of the other carbohydrate abnormalities, most likely diabetes mellitus (4).

Galactosemia

Of a much more severe nature are the inherited disorders associated with the improper metabolism of galactose in the newborn infant. As of this time two such disorders have been documented. It has recently been suggested that these be designated as "transferase deficiency galactosemia and galactokinase deficiency galactosemia" (5).

The first and most frequently recognized of these disorders is due to a deficiency of galactose-1-phosphate uridyl transferase (6). This disorder, if untreated, is associated with failure to thrive, hepatosplenomegaly liver disease, cataracts and mental retardation (5,7).

More recently, a second group of patients have been discovered who have galactosemia due to a deficiency of galactokinase (8). The clinical consequences of this form of the disease appears to be milder than the more classical galactosemia, with cataracts being the major finding (8,9).

While the ultimate diagnosis of either of these disorders rests upon the demonstration of the specific enzyme defect, the finding of excessive amounts of urinary galactose still remains the method by which many patients are first identified. The presence of a positive test for reducing substances in urine which yields a negative or weaker test result with methods based on glucose oxidase (Clinistix*) should suggest the possibility of these two disorders.

As several other sugars can yield similar results, one should follow-up these tests with a technique which can specifically identify galactose. Examples of such tests include galactose oxidase impregnated paper (10), paper chromatography (11) or thin layer chromatography (12,13).

Fructosuria and Fructose Intolerance

Another sugar which yields a positive Benedict's test and a negative glucose oxidase result is fructose. In contrast to xylulose and galactose, however, this compound yields a positive Seliwanoff's test (14). Of the two well-documented inherited disturbances of fructose, one essential fructosuria is asymptomatic and generally considered to be harmless (15). Hereditary fructose intolerance (16) on the other hand, can be associated with failure to thrive, vomiting, cirrhosis, mental retardation and, if left to its natural course, often early death (15). The

---

*Trade mark of the Ames Corporation, Elkhart, Indiana.

clinical course, however, appears to depend to a large degree on the amount of fructose in the diet of the affected individual (17).

Following the ingestion of foods rich in fructose, individuals with either of these disorders will have fructosuria resulting in a positive Benedict's test result. As with the other metabolic disorders, the nature of the sugar should be identified by additional procedures such as the Seliwanoff test and thin layer chromatography.

Specificity

The ability to yield a positive Benedict test is not limited to sugars as certain noncarbohydrate compounds also yield a positive result; i.e. homogentisic acid (18). The reaction with homogentisic acid, however, has been described as atypical, being a peculiar greenish black color (18).

As the above discussion indicates the identification of a urine containing excessive amounts of a "reducing substance" is only the first step in the analysis of that sample. One can, for example, perform several, more specific screening tests, the results of which can often give important clues as to the identity of the unknown compound as shown in Table 8-2. As one often finds a mixture of carbohydrates in urine, it is highly recommended that the identity of the reducing substance be confirmed by either paper or thin layer chromatography (11,12).

This is especially important as it has been shown that a number of factors might effect the various screening tests in such a way as to lead to incorrect conclusions. Naganna et al. have, for example, found false negative reactions with the Clinistix strips in presence of a glucose solution containing a variety of compounds (19). These included ascorbic acid, bilirubin glucuronide, homogentisic acid, adrenaline and hydroquinone. False negative results during routine urine screening have also been found to be due to dipyone and meralluride as well as vitamin C (20). More rarely, false positive results have been obtained with this test, presumably due to contamination with either hydrogen peroxide or some strong oxidizing agent such as hypochlorite (21).

It is also true, of course, that "false positive" results with the Benedict's test are associated with a variety of medications and chemical compounds. These include, but are not limited to, streptomycin (22), large doses of penicillin (23), and Renografin (24).

Incorrect conclusions based on screening test results can also arise in any situation in which more than one type of carbohydrate is present in the urine. This is particularly troublesome when testing urine from newborns, as it is not infrequent that urinary carbohydrates are detectable in the neonatal period (25,26).

TABLE 8-2.—Results of Various Screening Tests With Carbohydrates Associated With Inborn Errors of Metabolism

Sugars	Benedicts	Clinistix†	Urigalax‡	Seliwanoff's
Glucose	+	+	−	−
Galactose	+	−	+	−
Fructose	+	−	−	+
Xylulose*	+	−	−	−

*A large number of other carbohydrate and noncarbohydrate reducing substances will yield similar results.

†Trademark of the Ames Corporation of Elkhart, Indiana.

‡Trademark of the AB KABI, Norden flychyszagen 53, Box 30017, Stockholm, Sweden (should be commercially available in the latter part of 1972).

This may also be a problem in a specific inherited defect at any age, as it may happen that more than one sugar is excreted secondary to the primary disorder. Cornblath, for example, has cited a case in which the diagnosis of galactosemia was initially overlooked as a result of the presence of both glucose and galactose in the urine of the affected individual (34).

Qualitative Test for Reducing Substances

Method of Benedict (1)

Reagents:

1. Benedict's reagent: Prepare in the following manner:

    a. Dissolve 17.3 Gm $CuSO_4 \cdot 5H_2O$ in 100 ml of hot water.

    b. Dissolve 173 Gm of sodium citrate and 100 Gm of anhydrous $NaCO_3$ in about 800 ml of water. Heat to dissolve.

    c. When the two solutions are cool pour the second solution into the first while stirring.

    d. Dilute to a final volume of 1 liter.

2. This reagent is available commercially from the Fisher Scientific Company, Pittsburgh, Pa.

3. This solution is stable at room temperature.

Procedure:

1. Place 1.0 ml of the Benedict's reagent in a clean test tube.

2. Add 0.1 ml (2 drops) of urine and mix well.

3. Place in a boiling water bath for 3 minutes.

4. Observe the color immediately.

Results:

1. The results are reported as indicated in Table 8-3.

2. The following sugars yield a positive result:

Glucose	Xylulose
Fructose	Maltose
Galactose	Arabinose
Lactose	Xylose
Ribose	

TABLE 8-3.—Benedict's Reaction With Various Concentrations of Glucose (1)

Sugar Concentration	Color Reaction	Result
100 mg/100 ml	Blue or green	0
250 mg/100 ml	Green with yellow ppt	1+
800 mg/100 ml	Yellow to olive	2+
1,400 mg/100 ml	Brown	3+
2,000 mg/100 ml	Orange to red	4+

Notes:

1. Similar results may be obtained with Clinitest Tablets available from the Ames Corporation of Elkhart, Indiana. This screening test will also react with any reducing substance.

Urinary Glucose Test Using the Clinistix

Reagents:

1. Clinistix, Ames Company, Inc., Elkhart, Indiana.

Method:

1. Dip test end of the strip in the urine sample and remove immediately.
2. Compare the color of the dipped end with the color chart exactly 10 seconds after wetting.

Results:

1. Consult the color chart on the bottle label.

Notes:

1. This test is based on the glucose oxidase reaction and is generally considered to be specific for glucose.
2. As this test will yield negative results with sugars other than glucose it should not be used as a general screening test for abnormalities of carbohydrate metabolism.
3. "False negative" results have been reported to occur in the presence of ascorbic acid, bilirubin glucuronide, homogentisic acid, adrenaline, hydroquinone, dipyone and meralluride (19,20).

Qualitative Test for Fructose in Urine

Seliwanoff's Test (14)

Reagents:

1. Seliwanoff's reagent: Dissolve 0.05 Gm of resorcinol in a mixture of 60 ml of water and 30 ml of concentrated HCl.
2. Ethanol, 95%.

Procedure:

1. Place 5 drops of urine in a test tube.
2. Add 5 ml of Seliwanoff's reagent.
3. Boil for 30 seconds.
4. Centrifuge for 10 minutes in a clinical centrifuge and pour off the supernatant.
5. Add 3 ml of the 95% ethanol to the precipitate and mix well.
6. Observe for color.

Results:

1. A bright red color will be produced in the presence of fructose.

2. The red precipitate must be soluble in the 95% ethanol to be considered positive for fructose.

Notes:

1. Do not boil for more than 30 seconds or a small amount of glucose, if present, may give a positive result.

2. Glucose in concentrations above 2 Gm/100 ml will yield a positive result.

3. A positive test result obtained in an alkaline urine should be repeated in a fresh specimen as fructose may be formed from glucose under these conditions.

References

1. Henry, R. J.: *Clinical Chemistry: Principles and Technics* (New York: Paul B. Hoeber, Inc., Harper & Row, Publishers, 1964).

2. Garrod, A. E.: The Croonian Lectures on inborn errors of metabolism, Lancet 2:217-220, 1908.

3. Wang, Y. M., and van Eys, J.: The enzymatic defect in essential pentosuria, New England J. Med. 282:892-896, 1970.

4. Hiatt, H. H.: Pentosuria in Stanbury, J. B., Wyngaarden, J. B., and Fredrickson, D. S. (eds.): *The Metabolic Basis of Inherited Disease* 3rd Ed. (New York: McGraw-Hill Book Company, Inc., 1972).

5. Segal, S.: Disorders of Galactose Metabolism in Stanbury, J. B., Wyngaarden, J. B., and Fredrickson, D. S. (eds.): *The Metabolic Basis of Inherited Disease* (3rd ed.; New York: McGraw-Hill Book Company, Inc., 1972).

6. Isselbacher, K. J., Anderson, E. P., Kurahashi, K., and Kalckar, H. M.: Congenital galactosemia, a single enzymatic block in galactose metabolism, Science 123:635, 1956.

7. Isselbacher, K. J.: Galactose metabolism and galactosemia, Am. J. Med. 26:715-723, 1959.

8. Gitzelmann, R.: Deficiency of erythrocyte galactokinase in a patient with galactose diabetes, Lancet 2:670-671, 1965.

9. Gitzelmann, R.: Hereditary galactokinase deficiency, a newly recognized cause of juvenile cataracts, Pediat. Res. 1:14-23, 1967.

10. Dahlqvist, A.: Test paper for galactose in urine, Scandinav. J. Clin. & Lab. Invest. 22:87-93, 1968.

11. Wright, S. W., Ulstrom, R. A., and Szewczak, J. T.: Studies on carbohydrates in the body fluids. I. Identification by means of paper chromatography. J. Dis. Childhood 93:173-181, 1957.

12. Scott, R. M.: *Clinical Analysis by Thin Layer Chromatography Techniques.* (Ann Arbor: Humphrey Science Publishers, Inc., 1969).

13. Pifferi, P. G.: Improved thin layer chromatographic separation of hexoses and pentoses using kieselgel G, Anal. Chem. 37:925, 1965.

14. White, W. L., and Frankel, S.: *Seiverd's Chemistry for Medical Technologists* (2nd ed.; St. Louis: C. V. Mosby Company, 1965).

15. Froesch, E. R.: Essential Fructosuria and Hereditary Fructose Intolerance in Stanbury, J. B., Wyngaarden, J. B., and Fredrickson, D. S. (eds.): *The Metabolic Basis of Inherited Disease* (3rd ed.; New York: McGraw-Hill Book Company, Inc., 1972).

16. Froesch, E. R., Prader, A., Labhart, A., Stuber, H. W., and Wolf, H. P.: Die Hereditäre Fructoseintoleranz, eine bisher nicht bekannte kongenitale Stoffwechselstörung, Schweiz. med. Wchnschr. 87:1168-1171, 1957.

17. Cornblath, M., Rosenthal, I. M., Reisner, S. H., Wybregt, S. H., and Crane, R. K.: Hereditary fructose intolerance, New England J. Med. 269:1271-1278, 1963.

18. Smith, H. P., and Smith, H. P., Jr.: Ochronsis. Report of two cases. Ann. Int. Med. 42:171-178, 1955.

19. Naganna, B., Rajamma, M., and Vasudeva Rao, K.: On the failure of enzyme paper strips to detect glucose in certain abnormal urines, Clin. chim. acta 17:219-221, 1967.

20. Gifford, H., and Bergerman, J.: Falsely negative enzyme paper tests for urinary glucose, J.A.M.A. 178:423-424, 1961.

21. Caraway, W. T.: Chemical and diagnostic specificity of laboratory tests, Am. J. Clin. Path. 37:445-464, 1962.

22. Neuberg, H. W.: Streptomycin as a cause of false-positive Benedict reaction for glycosuria, Am. J. Clin. Path. 24:245-246, 1954.

23. Whipple, R. L., and Bloom, W. L.: The occurrence of false positive tests for albumin and glucose in the urine during the course of massive penicillin therapy, J. Lab. & Clin. Med. 36:635-639, 1950.

24. Howell, R. R.: Unpublished data.

25. Haworth, J. C., and MacDonald, M. S.: Reducing sugars in the urine and blood of premature babies, Arch. Dis. Childhood 32:417-421, 1957.

26. Dahlqvist, A., and Svenningsen, N. W.: Galactose in the urine of newborn infants, J. Pediat. 75:454-462, 1969.

27. Renold, A. E., Stauffacher, W., and Cahill, G. F., Jr.: Diabetes Mellitus in Stanbury, J. B., Wyngaarden, J. B., and Fredrickson, D. S. (eds.): *The Metabolic Basis of Inherited Disease* (3rd ed.; New York: McGraw-Hill Book Company, Inc., 1972).

28. Krane, S. M.: Renal Glycosuria in Stanbury, J. B., Wyngaarden, J. B., and Fredrickson, D. S. (eds.): *The Metabolic Basis of Inherited Disease* (3rd ed.; New York: McGraw-Hill Book Company, Inc., 1972).

29. Lindquist, B., and Meeuwisse, G. W.: Chronic diarrhoea caused by monosaccharide malabsorption, Acta paediat. scandinav. 51:674-685, 1962.

30. Darling, S., Mortensen, O., and Søndergaard, G.: Lactosuria and amino-aciduria in infancy. A new inborn error of metabolism, Acta paediat. 49:281-290, 1960.

31. Gray, G. M.: Intestinal Disaccharidase Deficiencies and Glucose-Galactose Malabsorption in Stanbury, J. B., Wyngaarden, J. B., and Fredrickson, D. S. (eds.): *The Metabolic Basis of Inherited Disease* (3rd ed.; New York: McGraw-Hill Book Company, Inc., 1972).

32. McCune, D. J., Mason, H. H., and Clarke, H. T.: Intractable hypophosphatemic rickets with renal glycosuria and acidosis (the Fanconi syndrome), Am. J. Dis. Child. 65:81-146, 1943.

33. Schneider, J. A., and Seegmiller, J. E.: Cystinosis and the Fanconi Syndrome in Stanbury, J. B., Wyngaarden, J. B., and Fredrickson, D. S. (eds.): *The Metabolic Basis of Inherited Disease* (3rd ed.; New York: McGraw-Hill Book Company, Inc., 1972).

34. Cornblath, M.: Clinical Aspects in Hsia, D. Y. Y. (ed.): *Galactosemia* (Springfield, Ill.: Charles C Thomas, Publisher, 1969).

CHAPTER 9

Sulfite Oxidase Deficiency: Sulfite Paper Test

In 1967, Mudd et al. described a patient whose urine contained increased amounts of S-sulfo-L-cysteine, sulfite and thiosulfate (1). The clinical findings included severe neurological abnormalities, mental retardation and dislocated ocular lenses (2). Detailed biochemical studies provided evidence that these findings were the result of a marked deficiency of the enzyme sulfite oxidase which is required for the oxidation of sulfite to sulfate (1,2).

Ferrocyanide-Nitroprusside Test

In an attempt to gain additional knowledge regarding the incidence, mode of inheritance and clinical importance of this disorder, Kutter and Humbel have introduced a simple urinary screening test (3). This test utilizes Feigl's ferrocyanide-nitroprusside reaction as a means of detecting the increased concentrations of the urinary sulfite found in this disorder (4). This test is based on the formation of a pink-red color on paper strips impregnated with sodium nitroprusside, zinc salts and potassium cyanoferrate in the presence of increased urinary sulfite. This test paper is commercially available from Macherey, Nagel & Company, Düren, West Germany.

Evidence has been presented which indicates that this test should be a useful tool for detecting additional patients with sulfite oxidase deficiency. It must be pointed out, however, that as of this time we are unaware of any patient who has been diagnosed with this procedure and thus proof of its value as a screening test is still lacking. We include it here due to its simplicity as a useful adjunct in evaluating patients with dislocated lenses. We have screened a considerable number of patients without "false positive" results. Moreover, sulfite added to urine from normal individuals clearly yields a positive result.

Sulfite Screening Test

Method of Kutter and Humbel (3)

Reagents:

1. Sulfite Test Paper, Macherey, Nagel and Company, D-516 Düren, West Germany.*

Method:

1. Dip paper strip into a *fresh* urine specimen.
2. Observe color reaction after 10 seconds.

---

*This product is distributed in the U.S.A. by the Gallard-Schlesinger Chemical Mfg. Corp., Carle Place, New York.

Results:

1. Urine containing sulfite produces an intense pink color within 10 seconds (3).

2. Normal urine samples yield no more than a slight pink discoloration.

Notes:

1. Testing must be carried out only on freshly voided urine as sulfite is rapidly oxidized spontaneously in urine (3).

2. "False positive" results have been obtained by high concentrations of ketone bodies and urobilinogens (3).

3. A thin layer chromatography method for S-sulfo-L-cysteine has been presented (3). As this compound is also increased in this disorder this method can be used as an additional means of detecting patients with this abnormality. Moreover, this compound is more stable than sulfite.

References

1. Mudd, S. H., Irreverre, F., and Laster, L.: Sulfite oxidase deficiency in man. Demonstration of the enzymatic defect, Science 156:1599-1601, 1967.

2. Irreverre, F., Mudd, S. H., Heizer, W. D., and Laster, L.: Sulfite oxidase deficiency. Studies of a patient with mental retardation, dislocated ocular lenses, and abnormal urinary excretion of S-sulfo-L-cysteine, sulfite and thiosulfate, Biochem. Med. 1:187-217, 1967.

3. Kutter, D., and Humbel, R.: Screening for sulfite oxidase deficiency, Clin. chim. acta 24:211-214, 1969.

4. Feigl, F.: *Spot Tests in Inorganic Analysis* (5th ed.; New York: Elsevier Press, Inc., 1958).

CHAPTER 10

Hyperuricemia: Uric Acid to Creatinine Ratio

For many years elevated concentrations of serum uric acid have been known to occur in the gouty patient. In recent years it has become apparent that there are syndromes of interest to the geneticist, other than gout, that are regularly associated with hyperuricemia.

Lesch-Nyhan Syndrome

With the description of the Lesch-Nyhan syndrome which is associated with dramatic elevations of serum and urinary uric acid, there was a flourish of interest in the study of uric acid in children (1). The Lesch-Nyhan syndrome has since been shown to be associated with an X-linked deficiency of the enzyme hypoxanthine guanine phosphoribosyl transferase which is apparently responsible for the dramatic increase in uric acid production (2). Clinically, this syndrome is characterized by mental retardation, choreoathetosis and self-mutilation. As already noted it is also associated with an increased production of uric acid.

Hyperuricemia in Other Genetic Disorders

Nyhan et al. have also found an elevated uric acid to creatinine ratio in a patient with hyperuricemia, dysplastic teeth, lack of speech, autistic behavior and absence of tears (3). Evidence that this is a new syndrome is seen in the fact that this patient had normal levels of hypoxanthine guanine phosphoribosyl transferase. In addition, Dr. J. Edwin Seegmiller has reported that patients with glycogen storage disease also have abnormally high ratios, being in the same range as these found in patients with the Lesch-Nyhan syndrome (4).

Uric Acid to Creatinine Ratio

A variety of sophisticated techniques are available to assess uric acid production, for example, the quantitation of isotope incorporation into uric acid isolated from urine. Careful quantitation of 24-hour urinary uric acid concentration is also, however, in most instances a useful tool to assess excretion and, hence, production of uric acid.

Kaufman et al., recognizing the great difficulty in obtaining clean carefully measured 24-hour urine samples in infants and retarded children, introduced the idea of measuring urinary uric acid and relating this to urinary creatinine (5). This uric acid to creatinine ratio is a useful tool in assessing marked over production of uric acid. It is well known that urinary creatinine is not without problems as a reference compound, but it clearly serves as a useful reference in this regard. Although there are some variations in the urinary uric acid excretion, it has been shown that patients who have dramatic overproduction of uric acid have distinctly abnormal urinary uric acid to creatinine ratios.

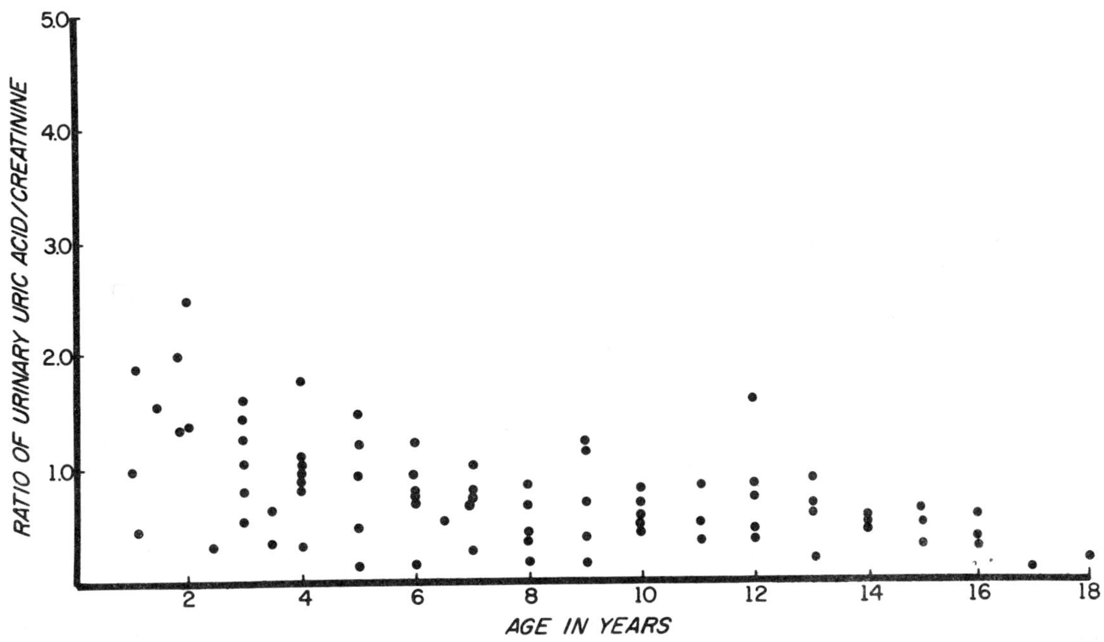

Fig. 10-1.—Range of urinary uric/creatinine ratios in random urine samples from patients in an institute for handicapped children.

The assessment of this ratio has been clearly shown to be useful in screening for patients with the Lesch-Nyhan syndrome. Kaufman et al. have published data on normal individuals as well as for patients affected by this syndrome (5). We have also found uric acid to creatinine ratios to be very much in the same range as those reported by this group (Fig. 10-1). As originally noted by Kaufman et al., the uric acid to creatinine ratio is inversely related to age (5). This group, thus, found that the mean value of the ratio which was 1.55 in the first week after birth had fallen to 0.61 by the 10th year of life.

Filter Paper Method

Recently, McInnes et al. have published a method which permits the collection of urine samples on filter paper which is then dried (6). The uric acid and creatinine are then eluted from a portion of the paper and quantitated. This method has also been found to be reliable in screening for uric acid abnormalities.

Methodology

The preferable method to assess uric acid concentrations in body fluids is to use the highly specific method employing purified uricase and measuring the conversion of uric acid to allantoin by this enzyme (7). For screening methods, however, it has been found that the usual techniques employed in the automated clinical chemistry laboratory are valid and useful in screening for abnormal uric acid to creatinine ratios.

Both the published results noted above (5,6) and our own data are based on the automated modifications of the Jaffe creatinine reaction and the Archibald uric acid method (8,9). As these are widely used standard procedures, it is recommended that one consult the Technicon Corporation method file N-30 (simultaneous creatinine and uric acid) for the detailed directions.

Alternatively, of course, one can utilize standard manual methods for the determination of uric acid (7) and creatinine (10) and express the results as a ratio. Evidence presented by Kaufman et al. has shown that differences between these various procedures are not significant for this type of study (5).

Uric Acid in Urine

Method of Liddle, Seegmiller and Laster (7)

Reagents:

1. Glycine buffer, 0.66M, pH 9.4: Dissolve 12.5 Gm of glycine and 2.2 Gm of sodium hydroxide in CO_2-free water and dilute to a final volume of 250 ml. Add 2 ml of chloroform as a preservative. Store in a refrigerator.

2. Dilute glycine buffer 0.1M: Dilute 7.6 ml of the 0.66M glycine buffer to a total volume of 50.0 ml with distilled water. Store in a refrigerator. Prepare a fresh solution weekly.

3. Purified uricase: Can be obtained commercially from the Worthington Biochemical Company, Freehold, New Jersey.

4. Stock uric acid standard, 30.0 mg/100 ml: Place 30.0 mg of uric acid in a 100-ml volumetric flask and add about 10 ml of hot distilled water. Dissolve the uric acid by adding 1.0 ml of a hot saturated solution of lithium carbonate. Dilute to a total volume of 100 ml with distilled water. *Store frozen.*

5. Dilute uric acid standard, 12 micrograms (μg) of uric acid per ml: Place 1.0 ml of the stock uric acid standard in a 25-ml volumetric flask and dilute this to a total volume of 25 ml with distilled water. Prepare fresh daily.

Sample:

1. Collect urine at room temperature. Collections can be preserved by using 4 ml of toluene as a preservative. Do not place in a refrigerator.

2. Dilute a 0.2 sample of the urine to a total volume of 10.0 ml with distilled water.

Procedure:

1. Allow all reagents except the enzyme solution to come to room temperature.

2. Standard and enzyme blank assay: Set up the following samples in quartz cuvettes for the determination of standard and enzyme blank values:

	Enzyme Blank	Standard #1	Standard #2
0.1M Glycine buffer	3.0 ml	2.0 ml	2.0 ml
Diluted uric acid standard	- - -	1.0 ml	1.0 ml

3. Mix and determine the O.D. values at 292 nm for each of the samples using water as a blank. This value is the initial O.D.

4. Add sufficient uricase to the enzyme blank and standards to produce a Δ O.D. in the standard cuvettes of at least 0.025 units per minute. (Normally this is 0.005-0.010 ml) of the enzyme solution.

5. Determine the O.D. of each sample at 25 and 30 minutes.

    a. The 25 and 30-minute readings are taken as a means of ascertaining if the reaction is complete.

    b. The 30-minute reading is the final O.D.

6. The above need be performed only once a day.

7. Calculations for standard.

    a. Δ O.D. Enzyme blank = final O.D. - initial O.D.

    b. Δ O.D. Standard cuvette = initial O.D. - final O.D.

    c. Δ O.D. Of uric acid standard = (a) + (b).

    d. Solve for X in the following equation:

$$X = \frac{\Delta \text{ O.D. of uric acid std.}}{12 \ \mu g} \times 3.01 \ ml^*$$

    e. X should equal 0.0745 O.D. units.

8. Urine uric acid assay: If the above standard assay is within acceptable limits procede to unknown urine samples. Set up the following in cuvettes.

	Biological Blank	Assay #1	Assay #2
0.1M Glycine buffer	2.0	2.0	2.0
Diluted unknown urine	1.0	1.0	1.0

9. Mix and determine the O.D. of each sample at 292 nm. This value is the initial O.D.

10. Add sufficient enzyme to assay #1 and #2 cuvettes (as determined from above), to complete reaction within 25 minutes. Do not add enzyme to the biological blank.

11. Determine O.D. at 25 and 30 minutes. If the reaction is not complete allow additional time for final reading.

12. Last reading (when the reaction is complete) is the final O.D.

13. Calculations for unknown urine samples.

    a. Δ O.D. assay cuvette = initial O.D. - final O.D.

    b. Δ O.D. biological blank = final O.D. - initial O.D.

    c. Δ O.D. enzyme blank = use enzyme blank value obtained during the standard determination above (step 7, part a).

    d. Δ O.D. uric acid = (a) + (b) + (c).

    e. Uric acid: mg/ml of urine $= \dfrac{\Delta \text{ O.D. uric acid}}{0.0745 \ \text{unit}/\mu g/ml} \times \dfrac{3.01 \ ml}{1 \ ml} \times \text{dilution}$

$$\times \frac{1 \ mg}{1,000 \ \mu g.}$$

---

*Assumes the addition of 0.01 ml of uricase.

Creatinine in Urine
Method Given by Henry (10)

Reagents:

1. Picric acid, 0.04M: Dissolve 2.034 Gm of reagent grade picric acid (containing 10%-12% H_2O) in distilled water and dilute to 200 ml final volume. Store in a dark bottle.

2. NaOH, 0.75N: Dissolve 6 Gm of NaOH in distilled water and dilute to a final volume of 200 ml.

3. HCl, 0.1N: Add 8.55 ml of concentrated acid to distilled water and dilute to 1,000 ml.

4. Stock creatinine standard, 150 mg/100 ml of 0.1N HCl: Dissolve 75 mg of creatinine in 0.1N HCl and dilute to a total of 50 ml.

5. Working creatinine standard, 1.5 mg/100 ml. Dilute stock standard 1:100 with distilled water. Prepare fresh each day.

Procedure:

1. Make 1:100 dilution of urine sample.

2. Set up test tubes in the following manner:

	Reagent	Standard	Unknown
H_2O	3.0 ml	1.0 ml	- - -
Working std.	- - -	2.0 ml	- - -
Urine 1:100	- - -	- - -	3.0 ml
Picric acid	1.0 ml	1.0 ml	1.0 ml
NaOH	1.0 ml	1.0 ml	1.0 ml

3. Mix well.

4. At the end of 20 minutes determine the O.D. at 520 nm using H_2O as a blank. Measure the time period with care.

5. Obtain corrected O.D. by subtracting reagent blank from std. or unknown reading.

Results:

1. $\dfrac{\text{O.D. unknown}}{\text{O.D. std.}} \times \text{conc. std.} \times \text{dil.} = \text{mg/ml}$

2. $\dfrac{\text{O.D. unknown}}{\text{O.D. std.}} \times \dfrac{0.010 \text{ mg}}{\text{ml}} \times 100 = \text{mg/ml}$

3. $\dfrac{\text{O.D. unknown}}{\text{O.D. std.}} = \text{mg of creatinine per ml of urine}$

Notes:

1. While it is sufficient to use one standard concentration for everyday use, one should prepare a standard curve at various times as an additional check. This can be done by diluting 0.5, 1.0, 2.0 and 3.0 of the working standard to a total of 3.0 ml and measuring as above.

References

1. Lesch, M., and Nyhan, W. L.: A familial disorder of uric acid metabolism and central nervous system function, Am. J. Med. 36:561-570, 1964.

2. Seegmiller, J. E., Rosenbloom, F. M., and Kelley, W. N.: An enzyme defect associated with a sex-linked human neurological disorder and excessive purine synthesis, Science 155:1682-1684, 1967.

3. Nyhan, W. L., James, J. A., Seberg, A. J., Sweetman, L., and Nelson, L. G.: A new disorder of purine metabolism with behavioral manifestations, J. Pediat. 74:20-27, 1969.

4. Howell, R. R.: Hyperuricemia in childhood, Fed. Proc. 27:1078-1082, 1968.

5. Kaufman, J. M., Greene, M. L., and Seegmiller, J. E.: Urine uric acid to creatinine ratio—a screening test for inherited disorders of purine metabolism, J. Pediat. 73: 583-592, 1968.

6. McInnes, R., Lamm, P., Clow, C. L., and Scriver, C. R.: A filter paper sampling method for the uric acid. Creatinine ratio in urine. Normal values in the newborn, Pediatrics 49:80-84, 1972.

7. Liddle, L., Seegmiller, J. E., and Laster, L.: The enzymatic spectrophotometric method for determination of uric acid, J. Lab. & Clin. Med. 54:903-913, 1959.

8. Nishi, H. H.: Determination of uric acid. An adaptation of the Archibald method on the Autoanalyzer, Clin. Chem. 13:12-18, 1967.

9. Chasson, A. L., Grady, H. J., and Stanley, M. A.: Determination of creatinine by means of automatic chemical analysis, Am. J. Clin. Path. 35:83-88, 1961.

10. Henry, R. J.: *Clinical Chemistry: Principles and Techniques* (New York: Paul B. Hoeber, Inc., Harper & Row, Publishers, 1964).

CHAPTER 11

Methylmalonic Aciduria: Thin Layer Chromatography Procedure

Methylmalonic aciduria is an inherited disorder of metabolism characterized by an inability to convert methylmalonyl coenzyme A to succinyl coenzyme A (1). This primary enzymatic dysfunction is in turn associated with a marked elevation in the concentration of methylmalonic acid in both blood and urine (2,3). The major clinical findings associated with these biochemical alterations in the neonatal period are lethargy, vomiting, ketoacidosis and a general failure to thrive (2,3).

If untreated, the prognosis for affected patients is very poor, being, in most cases, marked retardation in both growth and mental development and/or early death. Recent experience, however, indicates that the clinical course can be greatly improved by the use of either dietary control (4) or the administration of very high doses of vitamin B_{12} in some patients (5). It is, therefore, of great importance that newborns with this disorder be detected at the earliest possible opportunity.

Urinary Methylmalonic Acid

As affected patients excrete very large amounts of methylmalonic acid (2-6 Gm/day in contrast to a normal range of less than 5 mg), screening for this disorder can most easily be carried out by determining the urinary concentration of this compound (6).

Several methods for the quantitative determination of methylmalonic acid have been described. These include gas chromatography (7), paper chromatography combined with the titration of the isolated acid (8), anion-exchange chromatography combined with a colorimetric procedure (9), and the direct measurement of the compound in small amounts of urine without prior separation or purification steps (2,6). In addition, Carpenter *et al.* have developed a procedure which can be employed as a screening test to detect methylmalonic acid in urine saturated filter papers which are suitable for mailing through the postal service (10).

Chromatographic Screening Procedure

For screening purposes, however, we have found the application of the technique of thin layer chromatography, for both the detection and the identification of methylmalonic acid in the urine of patients, to be especially useful (11).

This procedure has been previously employed by others to detect and/or measure methylmalonic acid in urine of vitamin B_{12}-deficient patients suffering from pernicious anemia (12-15). For this purpose it has been found necessary, however, to extract and concentrate the methylmalonic acid from several milliliters of urine before the chromatography step (12-14).

We have found, however, that when this procedure is employed to detect patients with methylmalonic aciduria, one can analyze small amounts of untreated, unconcentrated urine directly without prior treatment. It is, therefore, possible to screen a large number of patients in very short periods of time at a minimum cost.

Thin Layer Chromatography of Methylmalonic Acid

Method of Thomas et al. (11)

Reagents and Equipment:

1. Chromatography medium: The following thin layer chromatography plates have been found to yield satisfactory results for the separation and identification of methylmalonic acid.

   a. Eastman Silica Gel Chromagram* Sheet #6061. These plates should be activated for at least 30 minutes at 100-110°C.

   b. Silica Gel G (E. Merck AG., Darmstadt, Germany). Mix 30 Gm of silica gel G with 60 ml of water by shaking vigorously in a 250-ml Erlenmeyer flask. Mix until the mixture is free of lumps. Spread the mixture on glass plates with an appropriate spreader, i.e. Desaga/Brinkmann apparatus for thin-layer chromatography, and allow to air dry for about 1/2 hour. Activate the silica gel by heating the plates in an oven at 100-110°C. for 1 hour before they are used for the chromatographic separation.

2. Chromatography solvent: A mixture consisting of ethanol-water-ammonium hydroxide (70:26:4) is used for either of the above chromatography mediums (14).

3. Diazo stain: The diazo stain is prepared by the method of Oberholzer et al. (2) as follows:

   a. HCl, 0.2N.: Dilute 1.7 ml HCl to a final volume of 100 ml with distilled water.

   b. p-Nitroaniline.[+] Dissolve 150 mg of p-nitroaniline† in 200 ml of 0.2N HCl. It might be necessary to warm the solution for approximately 5 minutes at 60-70°C.

   c. Sodium formate solution, pH 3.0: Prepare from formic acid and sodium hydroxide in the following manner:

      Add 17.4 ml of formic acid to distilled water and dilute to a final volume of 100 ml.
      Dissolve 16 Gm of sodium hydroxide in distilled water and dilute to a final volume of 100 ml.
      Combine 67 ml of the formic acid solution and 33 ml of the sodium hydroxide, mix and determine pH. If necessary adjust to a final pH of 3.0.

   d. Dilute formate solution: Add 5 ml of the sodium formate solution to 9 ml of 2N sodium hydroxide and 1 ml of water. Mix well.

   e. Sodium nitrite. Dissolve 330 mg of sodium nitrite in a total volume of 5 ml of distilled water. Prepare fresh just before needed.

---

*Registered trademark of the Eastman Kodak Company, Rochester, N.Y. 14650.
+This compound should be handled with care. May be fatal if absorbed through skin. Use only with adequate ventilation.
†Source: Eastman Kodak Company, Rochester, N.Y. 14650.

    f. Diazo reagent: Add 0.5 ml of the fresh sodium nitrite solution to 50 ml of the p-nitroaniline solution at room temperature. Mix and allow this mixture to stand for 2 minutes, add 7.5 ml of the dilute formate solution and mix well. Use immediately as a spray.

    g. Sodium hydroxide, 2N: Dissolve 8 Gm of NaOH in water and dilute to a final volume of 100 ml.

4. Methylmalonic acid* standard, 200 mg/100 ml: Dissolve 200 mg of methylmalonic acid in water made alkaline with 0.5 ml of 2N NaOH and dilute to a final volume of 100 ml with water.

Method:

1. Sample application: The methylmalonic acid standard and urine samples are spotted directly on the thin layer plate. No prior treatment or concentration of the samples is required. Samples are applied as follows:

    a. Small dots 2 cm apart and 2 cm from the bottom are made with a sharp pencil along one side of the plate.

    b. Apply a total of 10 μl of the standard and each urine sample to be analyzed in 5-μl amounts allowing the samples to dry between applications.

2. Chromatogram development: The chromatogram is developed in a standard T.L.C. chromatogram tank or Eastman Chromagram Developing Apparatus #6071 to which the chromatography solvent (ethanol-water-ammonium hydroxide, 70:26:4) has been previously added. This procedure is carried out in the following manner:

    a. Place the plate in the solvent tank at room temperature and allow the solvent to "run" for at least 3-5 hours. The Eastman Kodak Chromagram requires a longer "running time" with approximately 5 hours yielding satisfactory results.

    b. At the end of the run, remove the plate and allow it to air dry at room temperature for approximately 1 hour.

3. Visualization: The methylmalonic acid is detected by use of the diazo reagent. This step is carried out as follows:

    a. Carefully prepare a fresh solution of the diazo reagent and spray the chromatogram in a well-vented chemical hood. Allow the chromatogram to air dry.

    b. Spray the dry chromatogram with a solution of 2N NaOH.

Results:

1. Staining procedure: Methylmalonic acid forms a yellow spot when the chromatogram is sprayed with the fresh diazo reagent. Upon spraying with the 2N NaOH, this spot either remains yellow with the silica gel G plates or forms a green color when one employs the Eastman Chromagram sheet.

2. Sensitivity: This method will detect methylmalonic acid in amounts as small as 10 μg (100 mg/100 ml when using a 10-μl sample).

---

*Source: Aldrich Chemical Company, Inc., Milwaukee, Wis. 53233.

3. Findings: If one applies 10 μl of untreated, unconcentrated urine from normal persons, methylmalonic acid is not detectable by this method. In contrast, urine from patients with methylmalonic aciduria yields a rather strong reaction for methylmalonic acid with the same amount of urine. It is also possible that urine from patients with vitamin B_{12} deficiency will yield a positive test for this compound. A number of spots of various colors are found in urine from normal controls; however, none have the same color or R_f as methylmalonic acid.

References

1. Rosenberg, L. E., Lilljeqvist, A., and Hsia, Y. E.: Methylmalonic aciduria: Metabolic block localization and vitamin B_{12} dependency, Science 162:805-807, 1968.

2. Oberholzer, V. G., Levin, B., Burgess, E. A., and Young, W. F.: Methylmalonic aciduria. An inborn error of metabolism leading to chronic metabolic acidosis, Arch. Dis. Childhood 42:492-504, 1967.

3. Stokke, O., Eldjarn, L., Norum, K. R., Steen-Johnsen, J., and Halvorsen, S.: Methylmalonic acidemia: a new inborn error of metabolism which may cause fatal acidosis in the neonatal period, Scandinav. J. Clin. & Lab. Invest. 20:313-328, 1967.

4. Haslam, R.: Personal communication.

5. Hsia, Y. E., Lilljeqvist, A., and Rosenberg, L. E.: Vitamin B_{12}-dependent methylmalonic aciduria. Amino acid toxicity, long chain ketonuria, and protective effect of vitamin B_{12}, Pediatrics 46:497-507, 1970.

6. Giorgio, A. J., and Luhby, A. L.: A rapid screening test for the detection of congenital methylmalonic aciduria in infancy, Am. J. Clin. Path. 52:374-379, 1969.

7. Cox, E. V., and White, A. M.: Methylmalonic acid excretion: An index of vitamin B_{12} deficiency, Lancet 2:853-856, 1962.

8. Barness, L. A., Young, D., Mellman, W. J., Kahn, S. B., and Williams, W. J.: Methylmalonate excretion in a patient with pernicious anemia, New England J. Med. 268:144-146, 1963.

9. Giorgio, A. J., and Plaut, G. W. E.: A method for the colorimetric determination of urinary methylmalonic acid in pernicious anemia, J. Lab. & Clin. Med. 66:667-676, 1965.

10. Carpenter, D. G., Maddux, B. L., and Carter, C. H.: Screening test for methylmalonic aciduria utilizing urine-impregnated filter paper samples, J. Pediat. 80:108-110, 1972.

11. Thomas, G. H., Villarreal, M. L., Wells, C., and Bace, J.: Unpublished data.

12. Rosenberg, L. E., Lilljeqvist, A., and Hsia, Y. E.: Methylmalonic aciduria. An inborn error leading to metabolic acidosis, long-chain ketonuria and intermittent hyperglycinemia, New England J. Med. 278:1319-1322, 1968.

13. Dreyfus, P. M., and Dubé, V. E.: The rapid detection of methylmalonic acid in urine. A sensitive index of vitamin B_{12} deficiency, Clin. chim. acta 15:525-528, 1967.

14. Gompertz, D.: The measurement of urinary methylmalonic acid by a combination of thin-layer and gas chromatography, Clin. chim. acta 19:477-484, 1968.

15. Gutteridge, J. M. C., and Wright, E. B.: A simple and rapid thin layer chromatographic technique for the detection of methylmalonic acid in urine, Clin. chim. acta 27:289-291, 1970.

CHAPTER 12

Histidinemia: Histidine and Urocanic Acid Procedures

In 1961 Ghadimi, Partington and Hunter reported the findings on two siblings with a previously undescribed disorder of histidine metabolism (1). These children were detected as a result of a positive urinary ferric chloride test which suggested a diagnosis of phenylketonuria. Further study of the serum and urinary amino acids, however, showed these children to have normal phenylalanine concentrations which excluded the diagnosis of phenylketonuria. These studies did, however, also reveal dramatic elevations of histidine in both of these fluids. These were the first recognized persons with histidinemia.

Since that time, approximately 40 patients have been reported who have genetic defects in the metabolism of histidine. The enzymatic defect has been shown in these children to consist of a deficiency of the activity of histidase (histidine alpha-deaminase) with a resultant inability to convert histidine to urocanic acid, the first major step of its catabolic pathway (2).

Urinary Imidazolepyruvic Acid

As a result of this metabolic abnormality the concentration of L-histidine is greatly increased in various body fluids such as serum and urine. In addition, as the concentration of histidine increases, there is a marked increase in its conversion to imidazolepyruvic acid by the enzyme, histidine transaminase. It is this increased production of imidazolepyruvic acid which yields the stable blue-green color with the ferric chloride test (see Chapter 3).

As noted in Chapter 3, however, there is often a normal delayed maturation of the histidine transaminase (3). As a consequence there may be a period of several weeks to months before there is sufficient urinary imidazolepyruvic acid present to yield a positive ferric chloride result in a newborn patient with histidinemia (3). It is, therefore, quite useful to have additional screening tests available for the detection of this disorder which are not based on the formation of imidazolepyruvic acid.

Urinary Histidine

Gerber and Gerber have described one such method which is based on the semiquantitative determination of the histidine present in random or 24-hour samples of urine (4). This test is based on the ability of excessive amounts of histidine to inhibit the formation of a blue color by the reaction between copper and bis-cyclohexanone oxaldihydrazone. Thus, by the appropriate dilutions of urine samples one may obtain an estimation of the urinary histidine concentration and thereby detect patients with histidinuria.

Urocanic Acid Deficiency

Alternatively, one can determine the concentration of urocanic acid in either sweat or stratum corneum as a means of detecting histidinemic subjects (5,6). Normally histidine is converted to urocanic acid, and as urocanase is absent in normal skin, the urocanic acid accumulates in the stratum corneum. Histidinemic subjects, on the other hand, are unable to form urocanic acid and, therefore this compound is absent from the stratum corneum of affected individuals.

When sweat collections are made using pilocarpine iontophoresis or thermal stimulation, urocanic acid is leached out of the skin and appears in the sweat collection. Thus, by measuring the urocanic acid in sweat one can detect patients with this disorder (5).

Urocanic Acid Assay

Sweat, collected by one of the above methods, is diluted in 0.2M sodium phosphate buffer, pH 7.5, and examined spectrophotometrically. This is done by scanning between 220 and 320 nm using 0.2M sodium phosphate as a blank. The presence of urocanic acid is indicated by a strong absorption at 277 nm under these conditions. As urocanic acid is normally the major compound in sweat which absorbs light at this wavelength, the absence of absorption can be taken as good evidence for the absence of this compound. It is important, however, that the filter paper in which the sweat is collected be in contact with the skin during the collection period. This appears to be due to the fact that normal sweat does not contain urocanic acid when direct skin contact is avoided as is shown in Figure 12-1.

Fig. 12-1.—Samples of sweat, as well as authentic urocanic acid, were diluted in 0.1M potassium phosphate buffer, pH 7.4, for spectral studies. Dilutions, depending on the amount of sweat obtained, were between 1:5 and 1:10. Light path = 1 cm. It is apparent that normal sweat does not contain urocanic acid when skin contact is avoided.

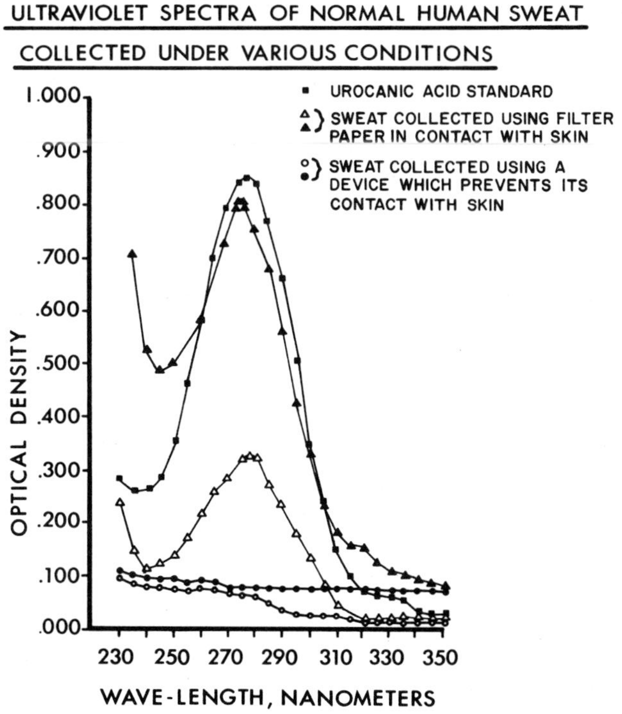

In another screening test, originally published by Levy et al., one simply assays directly for the presence or absence of urocanic acid in homogenates of small amounts of skin clippings (6). In this method, however, the presence of urocanic acid is demonstrated by the use of paper chromatography followed by staining with Pauly's reagent.

As urinary screening test of Gerber and Gerber appears likely to yield, on occasion, false positive results for the presence of increased histidine, we feel that the tests based on the presence or absence of urocanic acid are more reliable.

Histidine in Urine

Method of Gerber and Gerber (4,7)

Reagents:

1. Tris buffer, pH 7.4, 0.1M: Dissolve 12.1 Gm of tris (hydroxymethyl) aminomethane in a total of 500 ml of distilled water. To 125 ml of this solution add 103.5 ml of 0.2N HCl and dilute the mixture to a final volume of 250 ml with water. Determine the pH and if necessary adjust to 7.4.

2. Sodium citrate, 1M: Dissolve 29.41 Gm of sodium citrate dihydrate in water and dilute to a final volume of 100 ml.

3. Copper sulfate, 0.5%: Dissolve 0.5 Gm of $CuSO_4 \cdot 5H_2O$ in water and dilute to a final volume of 100 ml.

4. Copper-buffer solution: Add 0.5 ml of the 0.5% copper sulfate to 250 ml of the 0.1M tris buffer and 1.6 ml of the 1M sodium citrate solution.

5. Cuprizone solution: Dissolve 10.0 mg of bis-cyclohexanone oxaldihydrazone* [oxalic acid bis (cyclohexylidenehydrazine)] in a mixture of 5 ml of ethyl alcohol and 5 ml H_2O.

Procedure:

1. Place 4 ml of the copper-buffer solution in a clean test tube.

2. Add 0.2 ml of urine to the above solution and mix well.

3. Add 0.25 ml of the cuprizone reagent and mix.

4. Allow to stand for five minutes at room temperature.

5. Observe color of the solution.

Results:

1. A positive result is a colorless or very slightly blue solution. This result indicates the presence of increased amounts of histidine.

2. A negative result is a blue colored solution. This finding indicates a normal urinary histidine concentration.

---

*Source: G. Frederick Smith Chemical Company, Columbus, Ohio.

Notes:

1. This test measures both urinary histidine and 3-methylhistidine.

2. A negative result (blue solution) is reported to indicate that the urinary histidine is below 60 mg/ml (4).

3. A positive result (colorless or slightly blue) is reported to indicate that the histidine concentration is above 20 mg/100 ml (4).

4. Due to the overlap between negative and positive results (presumably due to interfering compounds), it has been suggested that a 1:1 dilution be made on any positive urine and that the test be repeated (4).

Urocanic Acid in Skin Clippings

Method of Levy et al. (6)

Reagents and Materials:

1. Ammonium hydroxide, concentrated.

2. Urocanic acid standard, 100 mg/100 ml: Dissolve 10 mg of urocanic acid (4-imidazole acrylic acid) in a total of 10 ml of distilled water.

3. Histidine standard, 100 mg/100 ml: Dissolve 10 mg of DL-Histidine HCl in a total of 10 ml of distilled water.

4. Chromatography solvent: Butanol, acetic acid and water: 36:9:15.

5. Hydrochloric acid, 1N: Add 8.33 ml of concentrated HCl to about 75 ml of distilled water and dilute to a final volume of 100 ml.

6. Sulfanilic acid, 1% in HCl: Dissolve 500 mg of sulfanilic acid in a total of 50 ml of 1N HCl.

7. Sodium nitrite solution, 5%: Dissolve 2.5 Gm of sodium nitrite ($NaNO_2$) in water and dilute to a final volume of 50 ml.

8. Pauly's reagent: Mix 5 ml of the 1% sulfanilic acid with 25 ml of the 5% sodium nitrite just before it is required for the spray..

9. Sodium carbonate, 5%: Dissolve 5 Gm of sodium carbonate in a total of 100 ml of water.

10. Whatman 3-MM chromatography paper: 9 X 9 in square.

11. Chromatography jar.

Procedure:

1. Remove small clippings from the stratum corneum of the finger tips or toes with ordinary fingernail clippers (8).

2. Place from 1-5 mg of these clippings in concentrated ammonium hydroxide (1 drop of ammonium hydroxide for each milligram of stratum corneum) in a small mortar.

3. Homogenize this mixture with a small pestle.

4. Apply the above solution on a small dot 2 cm from the edge of one end of the chromatography paper.

5. Apply 10 µl of the urocanic acid and histidine standards to separate spots in a similar manner along the same edge of the chromatography paper.

Chromatography:

1. Develop the chromatogram at room temperature in an ascending system for eight hours in the butanol, acetic acid and water solvent.

2. At the end of the run, remove the chromatogram and air dry at room temperature.

3. After drying, the chromatogram is sprayed with Pauly's reagent.

4. Immediately after spraying with Pauly's reagent, spray with a 5% solution of sodium carbonate.

Notes:

1. Upon spraying, both histidine and urocanic acid develop an immediate red color.

2. The spray reagents should be refrigerated and used while cold.

3. Quantities as small as 0.2 µg of urocanic acid can be visualized by this method.

4. Normal controls were found to have 1-5 µg of urocanic acid per mg of skin while no urocanic acid could be detected in skin samples from histidinemic subjects (6).

References

1. Ghadimi, H., Partington, M. W., and Hunter, A.: A familial disturbance of histidine metabolism, New England J. Med. 265:221-224, 1961.

2. LaDu, B. N., Howell, R. R., Jacoby, G. A., Seegmiller, J. E., and Zannoni, V. G.: The enzymatic defect in histidinemia, Biochem. Biophys. Res. Commun. 7:398-402, 1962.

3. Levy, H. L., Madigan, P. M., and Peneva, P.: Evidence for delayed histidine transamination in neonates with histidinemia, Pediatrics 47:128-131, 1971.

4. Gerber, M. G., and Gerber, D. A.: A simple screening test for histidinuria, Pediatrics 43:40-43, 1969.

5. Rosenblatt, D., Mohyuddin, F., and Scriver, C. R.: Histidinemia discovered by urine screening after renal transplantation, Pediatrics 46:47-53, 1970.

6. Levy, H. L., Baden, H. P., and Shih, V. E.: A simple indirect method of detecting the enzyme defect in histidinemia, J. Pediat. 75:1056-1058, 1969.

7. Duncan, I. W.: Urinary screening tests to detect metabolic disorders, Alaska Med. Sept.: 86-89, 1970.

8. Howell, R. R.: Diagnostic procedures for Genetic Metabolic Defects in Bartalos, M. (ed.): *Genetics in Medical Practice* (Philadelphia: Lea & Febiger, 1968).

CHAPTER 13

Metachromatic Leukodystrophy: Urinary Arylsulfatase A Test

In 1947 Huggins and Smith reported that human urine contained arylsulfatase activity (1). It has since been found, using electrophoretic techniques, that this activity is due to the presence of at least two distinct enzymes (2). It was found, moreover, that these enzymes are identical with the arylsulfatases A and B of human tissue. Due to the fact that these two enzymes are inhibited by different inorganic salts, it is now possible to determine the two activities independently without prior separation (3).

There has been considerable interest in these enzymes as there have been several reports of increased levels of urinary sulfatase activity in persons with various types of tumors. This area of investigation has been reviewed by Dzialoszyński and Gniot-Szulzycka (4).

Enzyme Defect

In the course of studying nine enzyme systems in tissue samples, Austin et al. observed that samples from metachromatic leukodystrophy patients showed consistently low arylsulfatase A activity (5). Evidence has been obtained by Mehl and Jatzkewitz (6) that the arylsulfatase A enzyme also has cerebroside sulfatase activity and that as a consequence of this enzymatic defect the conversion of cerebroside sulfate to cerebroside is blocked. As the reduced sulfatase A activity is also found in the urine of patients with metachromatic leukodystrophy (7), a urinary screening test for arylsulfatase A activity was introduced by Austin and his associates (8). This has been found by others to be a reliable test to screen for patients with metachromatic leukodystrophy (9).

Screening Test of Austin et al.

In this screening test arylsulfatases A and B are precipitated from urine with ammonium sulfate to concentrate the enzyme and remove at least some inhibitors. The precipitated enzymes are then dissolved and incubated with nitrocatechol sulfate (2-hydroxy-5-nitrophenyl sulfate) under conditions which measure only the arylsulfatase A activity. The enzyme removes the sulfate group from the nitrocatechol sulfate resulting in free nitrocatechol. This compound is soluble in ether when it is in a nonionic form (acid pH). The free nitrocatechol in the ionic form (basic pH) develops a burgundy red color and is also concentrated in the NaOH due to its solubility properties (nitrocatechol sulfate does not form color in the NaOH). The amount of red color is thus a direct indication of the amount of enzyme present in the urine.

TABLE 13-1.–Arylsulfatase A Activity in Urine Samples Obtained from Patients with Metachromatic Leukodystrophy (MLD) and Control Patients

	Condition	Number of Specimens	Units of Arylsulfatase A Activity[*]	
			Average	Range
A.	Late infantile MLD			
	Patient No. 1	13	0.4	0.0 – 2.4
	Patient No. 2	5	0.2	0.0 – 0.7
	Patient No. 3	4	0.3	0.0 – 0.5
B.	Adult MLD			
	Patient No. 1	4	1.3	0.4 – 2.0
	Patient No. 2	6	1.5	0.3 – 2.0
	Patient No. 3	2	1.2	0.8 – 1.5
C.	Control group			
	Patients (160)	206	20.5	1.3[†] – 162.8

[*]1 Unit equals 1 μg of nitrocatechol released by 1 ml of urine in 1 hour at 37°C.

[†]Only 3 control patients of the 160 (1.9%) had an average enzyme value equal to or below the highest single value (2.4 units) ever observed in the patients with MLD.

From Thomas, G. H., and Howell, R. R.: Arylsulfatase A activity in human urine. Quantitative studies on patients with lysosomal disorders including metachromatic leukodystrophy, Clin. chim. acta 36:99-103, 1972.

Screening Test of Baum et al.

One of the difficulties encountered with the screening test of Austin (8) is obtaining the required 100 ml urine specimen needed for the test. This is especially true in a pediatric service laboratory. One can overcome this difficulty, however, by using the assay of Baum et al. (3). This assay method requires only 4 ml of urine and yields quantitative results. It has, moreover, been found to be quite useful as an aid in detecting patients with metachromatic leukodystrophy (Table 13-1) (10). We, therefore, prefer the method of Baum et al. (3), which is not only technically quite simple to perform, but also requires much less technician time than the Austin screening test.

Diagnostic Follow-up

While the test of Baum et al. (3) has been found to be a very useful means of identifying patients with metachromatic leukodystrophy (10), it is still, nevertheless, strongly recommended that this test be used only as a screening test and never as the basis of a positive diagnosis. If the average activity of at least three separate, carefully collected, fresh urine samples is below 5.0 units, additional clinical and/or enzymatic data should be acquired to confirm or to rule out the working diagnosis of metachromatic leukodystrophy. Several enzymatic techniques using leukocytes or fibroblasts have been described and have been found quite useful for this purpose (11, 12).

Screening Test for Arylsulfatase A Deficiency in Metachromatic Leukodystrophy

Method of Austin et al. (8)

Reagents:

1. Ammonium sulfate: $(NH_4)_2 SO_4$, granular, pyridine free.

2. Acetic acid, 0.5M: Dilute 2.9 ml of 99% acetic acid to 100 ml with distilled water.

3. Sodium acetate, 0.5M: Dissolve 6.80 Gm of $NaC_2H_3O_2 \cdot 3H_2O$ in distilled water and dilute to 100 ml.

4. Acetic acid-sodium acetate buffer pH 5.0, 0.5M: Add 30 ml of 0.5M acetic acid to 70 ml of 0.5M sodium acetate. Adjust final pH to exactly 5.0.

5. Acetic acid-sodium acetate buffer pH 5.0, 0.1M: Dilute 20 ml of 0.5M acetate buffer to 100 ml with water. Adjust pH to 5.0.

6. Sulfatase A reagent: Dissolve 82.5 mg of p-nitrocatechol sulfate,* 5.6 mg of sodium pyrophosphate $(Na_4P_2O_7 \cdot 10H_2O)$ and 2.5 Gm of NaCl in a small amount of 0.5 acetic acid-sodium acetate buffer and dilute to 25 ml with the buffer.

7. HCl, 1N stock: Dilute 8.55 ml of concentrated HCl to 100 ml with distilled water.

8. HCl, 0.01N: Make a 1:100 dilution of the 1N HCl stock solution.

9. HCl, 0.05N: Make a 1:20 dilution of the 1N HCl stock solution.

10. Acidified ether solution: Shake 5 volumes of ether with 1 volume of .05N HCl, allow phases to separate and remove upper ether phase for later use.

11. NaOH, 6N: Dissolve 24 Gm of NaOH in water and dilute to 100 ml.

Procedure:

1. Check pH of a *fresh* sample of urine; if alkaline, add 0.1M acetic acid until pH is between 5 and 6.

2. Filter 100 ml of urine through filter paper which has been previously moistened with a small amount of distilled water.

3. Slowly add 47.3 Gm of ammonium sulfate to the filtered urine. Gently swirl until the crystals are dissolved.

4. Centrifuge for 10-15 minutes at 6,000 × g. (Centrifuge longer if supernatant is not clear.)

5. Remove supernatant taking care not to disturb pellet.

6. Dissolve pellet in 1-2 ml of acetate buffer (0.1M, pH 5.0) at room temperature. Determine pH with paper and add more buffer to achieve pH 5.0 if required.

7. Add 1.5 ml of the sulfatase reagent solution to the above solution. Prepare a blank by using 1.5 ml of 0.1M acetate buffer in place of "pellet solution." Normal urines should also be run with each test.

8. Incubate these solutions at 75°C. for 10 minutes.

9. At end of 10 minutes add 10 ml of .01N HCl to each tube. Invert 5 times and allow to stand for 5 minutes. Add 5 ml of acidified ether to each tube. Invert gently 5 times (carefully release pressure after each inversion). Remove ether from each tube and place in clean tubes.

10. Repeat extraction 2 additional times and pool ether extracts (15 ml total in each tube).

11. Remove aqueous phase, if any, from bottom of each tube of pooled ether extract.

12. Add 1 large drop of 6 N NaOH, mix by tapping the tube.

13. Allow to stand 5 minutes and then record color in bottom of tube.

---

*Source: Sigma Chemical Company, 3500 Dekalb St., St. Louis, Mo. 63118.

Results:

μg of 4-Nitrocatechol	Color	Grade
0	Pale yellow to yellow-orange	0
1	Definite orange-pink	1+
5	Red-orange	2+
10	Red	3+
20	Dark red	4+

Patients with metachromatic leukodystrophy normally only show a faint trace (0 to 1+ of sulfatase A activity, while controls normally show a definite 1+ to 4+ result).

Quantitative Method for Arylsulfatase A in Urine

Method of Baum et al. (3,10)

Reagents:

1. Acetic acid, 0.5M: Dilute 2.9 ml of 99% acetic acid to 100 ml with distilled water.

2. Sodium acetate, 0.5M: Dissolve 6.80 Gm of Na $C_2H_3O_2 \cdot 3H_2O$ in distilled water and dilute to 100 ml.

3. Acetic acid-sodium acetate buffer, pH 5.0, 0.5M: Add 30 ml of 0.5M acetic acid to 70 ml of 0.5M sodium acetate. Adjust the final pH to exactly 5.0.

4. Sodium hydroxide, 1N: Dissolve 4.0 Gm of NaOH in water and dilute to 100 ml.

5. Sulfatase A reagent: Dissolve 82.5 mg of p-nitrocatechol sulfate,* 5.6 mg of sodium pyrophosphate ($Na_4P_2O_7 \cdot 10H_2O$) and 2.5 Gm of NaCl in a small amount of 0.5M acetic acid-sodium acetate buffer and dilute to 25 ml with the same buffer.

Method:

1. Obtain fresh urine.

2. Place in container which is kept on ice.

3. Adjust pH of the urine to 6.0-6.3. Measurement of pH can be done with pH paper.

4. Filter or centrifuge for 5 minutes in a refrigerated centrifuge.

5. Place 4 ml of urine in dialysis tubing.

6. Place tubing in 1 liter of precooled distilled water (0-5°C) and dialyse with stirring for at least 18 hours. Change the water at least 3 times during this period using precooled distilled water.

7. Remove sample and adjust to a final volume of 6 ml with distilled water.

8. Prepare incubation mixture in following manner:

	Tube 1	Tube 2	Tube 3
Urine	.5	.5	- - -
Sulfatase A	.5	- - -	.5

9. Mix well and incubate for one hour at 37.5°C in a water bath.

---

*Source: Sigma Chemical Company, 3500 Dekalb St., St. Louis, Mo. 63118.

10. At end of 1 hour add contents of tube #3 to tube #2. Add 1.5 ml of 1N NaOH to tubes 1 and 2. Mix well.

11. Measure O.D. of each tube against water blank at 515 nm.

Results:

1. Calculations are made in following manner:

155 = Molecular weight of 4-nitrocatechol.
12,400 = Molar extinction coefficient.
10^6 = Conversion of Grams to micrograms.
10^3 = Conversion of liter to milliliter.
2.5 = Final volume.
0.5 = Original volume of urine.
6/4 = Dilution due to dialysis treatment.

$$\frac{(\text{O.D. Sample} - \text{O.D. blank}) \times 155 \times 10^6 \times 2.5 \times 6}{12,400 \times 10^3 \times 0.5 \times 4} = \mu g \text{ of 4-nitrocatechol}$$

released in 1 hour by 1 ml of urine at 37°C.

or (O.D. Sample - O.D. blank) $\times$ 94 = μg 4-nitrocatechol released in 1 hour by 1 ml of urine at 37°C.

Notes:

1. It has been found that urine from normal controls may have a low value on isolated occasions. It is, therefore, suggested that several samples be assayed before a decision is made regarding the diagnosis of metachromatic leukodystrophy.

2. The results of our experience with this test are given in Table 13-1.

3. Our experience indicates that the results can be divided as follows:

0-2.9 units, low
3-5 units, borderline
> 5 units, normal

References

1. Huggins, C., and Smith, D. R.: Chromogenic substrates. III. p-Nitrophenyl sulfate as a substrate for the assay of phenolsulfatase activity, J. Biol. Chem. 170:391-398, 1947.

2. Dodgson, K. S., and Spencer, B.: The occurrence of arylsulphatases A and B in human urine, Clin. chim. acta 1:478-480, 1956.

3. Baum, H., Dodgson, K. S., and Spencer, B.: The assay of arylsulphatases A and B in human urine, Clin. chim. acta 4:453-455, 1959.

4. Dzialoszyński, L. M., and Gniot-Szulzycka J.: Some clinical aspects of arylsulphatase activity, Clin. chim. acta 15:381-386, 1967.

5. Austin, J. H., Balasubramanian, A. S., Pattabiraman, T. N., Saraswathi, S., Basu, D. K., and Bachhawat, B. K.: A controlled study of enzymic activities in three human disorders of glycolipid metabolism, J. Neurochem. 10:805-816, 1963.

6. Mehl, E., and Jatzkewitz, H.: Evidence for the genetic block in metachromatic leucodystrophy (ML), Biochem. Biophys. Res. Commun. 19:407-411, 1965.

7. Austin, J., McAfee, D., and Shearer, L.: Metachromatic form of diffuse cerebral sclerosis. IV. Low sulfatase activity in the urine of nine living patients with metachromatic leukodystrophy (MLD), Arch. Neurol. 12:447-455, 1965.

8. Austin, J., Armstrong, D., Shearer, L., and McAfee, D.: Metachromatic form of diffuse cerebral sclerosis. VI. A rapid test for the sulfatase A deficiency in metachromatic leukodystrophy (MLD) urine, Arch. Neurol. 14:259-269, 1966.

9. Greene, H., Hug, G., and Schubert, W. K.: Arylsulfatase A in the urine and metachromatic leukodystrophy, J. Pediat. 71:709-711, 1967.

10. Thomas, G. H., and Howell, R. R.: Arylsulfatase A activity in human urine. Quantitative studies on patients with lysosomal disorders including metachromatic leukodystrophy, Clin. chim. acta 36:99-103, 1972.

11. Percy, A. K., and Brady, R. O.: Metachromatic leukodystrophy: Diagnosis with samples of venous blood, Science 161:594-595, 1968.

12. Kaback, M. M., and Howell, R. R.: Infantile metachromatic leukodystrophy. Heterozygote detection in skin fibroblasts and possible applications to intrauterine diagnosis, New England J. Med. 282:1336-1340, 1970.

Generalized Gangliosidosis: Urinary β-Galactosidase Test

In 1967, Sacrez et al. (1) reported that Hers had found a deficiency of β-D-galactosidase in the liver of a patient with generalized gangliosidosis (GM_1 gangliosidosis, Landing's disease, Tay-Sachs disease with visceral involvement). This observation has since been confirmed and expanded by other investigators (2,3). The clinical and biochemical aspects of this disease have been reviewed in detail by O'Brien (4).

Biochemical Defect

Generalized gangliosidosis is inherited in an autosomal recessive fashion and is characterized, in part, by the accumulation of large quantities of GM_1 ganglioside. This abnormal accumulation is found primarily in the brain and visceral organs. GM_1 ganglioside has been shown to have the following structure:

Galactosyl - (1 → 3) - N-acetylgalactosaminyl - (1 → 4) -

[(2 → 3) - N-acetylneuraminyl] - galactosyl - (1 → 4) -

glucosyl - (1 → 1) - [2 - N-acyl] - sphingosine.

This compound is normally hydrolyzed by a group of glycosidases located in the lysosome. The initial step in the degradative pathway involves the conversion of GM_1 to GM_2 ganglioside by the cleavage of the β-linked terminal galactose from the GM_1 by a β-D-galactosidase (5). In the absence of this activity the conversion does not take place.

Enzyme Assay

A number of artificial substrates such as p-nitrophenyl-β-galactopyranoside, o-nitrophenyl-β-D-galactopyranoside and 4-methylumbelliferyl-β-D-galactopyranoside are also cleaved by the galactosidase. In contrast to the natural substrate(s), these artificial compounds allow for enzyme assays which are both simple and easy to perform in any laboratory.

The assay of β-D-galactosidase activity in various human tissues and fluids utilizing artificial substrates permits the detection of patients with generalized gangliosidosis. Methods for the assay of β-galactosidase using artificial substrates have been employed by a number of investigators, using various tissue preparations (2), skin fibroblasts (3), amniotic fluid fibroblasts (3), plasma (6) and urine (7).

Urinary Screening Test

We have found the assay of β-galactosidase in urine, while presenting certain problems, to be useful as a screening method for this disorder. The usefulness of the urinary assay arises from the fact that the concentration of β-D-galactosidase in urine appears to reflect, to a large degree, the enzyme activity of various tissues and organs from which it arises. Thomas has, therefore, only found consistently low β-galactosidase levels in urine from a patient with generalized gangliosidosis. In contrast, a large number of controls were found to have much higher levels of enzyme activity (7).

It should be pointed out that variations in random factors, such as pH, age, dilution and collection methods of the urine, might on occasion result in low enzyme activities in urine from normal individuals. While it is evident that repeated assays on several urine samples from a single patient should correct for random factors, it has been found that the simultaneous assay of a second lysosomal enzyme, subject to the same variable factors, serves as a better control. By relating the two enzyme activities to each other as a ratio, it is possible to separate quite clearly the patient from the control group. In the method given here, N-acetyl-β-D-glucosaminidase is used as the control enzyme. The results are then given as a ratio of N-acetyl-β-D-glucosaminidase to β-D-galactosidase.

Screening Test for β-D-Galactosidase Deficiency in Generalized Gangliosidosis

Method of Thomas (7)

Reagents:

1. Acetate buffer, 0.5M, pH 5.0: Dilute 2.86 ml of 99% acetic acid to a final volume of 100 ml. Add 29.6 ml of this solution to 70.4 ml of a 0.5M solution of sodium acetate (6.8 Gm of $C_2H_3O_2Na \cdot 3H_2O$ in distilled water made to a final volume of 100 ml). Adjust to a final pH of 5.0.

2. p-Nitrophenyl-β-D-galactoside, 0.005M in 0.5M acetate buffer pH 5.0: Dissolve 3.01 mg of p-nitrophenyl-β-D-galactoside monohydrate* in a small volume of acetate buffer and dilute to a final volume of 2.0 ml.

3. p-Nitrophenyl-N-acetyl-β-D-glucosaminide, 0.00375M in 0.5M acetate buffer, pH 5.0: Dissolve 2.56 mg of p-nitrophenyl-N-acetyl-β-D-glucosaminide[+] in a small volume of acetate buffer and dilute to a final volume of 2.0 ml.

4. Glycine buffer, 0.25M, pH 10.0: Dissolve 1.88 Gm of glycine and 2.65 Gm of sodium carbonate in distilled water and dilute to a final volume of 100 ml.

5. p-Nitrophenol standard stock solution, 1,000 mμ mols/ml: Dissolve 13.9 mg of p-nitrophenol in distilled water and dilute to a final volume of 100 ml.

Method:

1. Obtain a fresh random sample of urine.

2. Filter through Whatman No. 1 filter paper into a chilled beaker on ice.

3. Place exactly 10 ml of the filtered sample in 5/8″ presoaked dialysis tubing which retains material with a molecular weight of 12,000 and above.

---

*Source: Sigma Chemical Company, 3500 Dekalb St., St. Louis, Mo. 63118.
[+]Source: Pierce Chemical Company, P.O. Box 117, Rockford, Ill. 61105.

4. Dialyze for at least 15 hours at 3-6°C. against several changes of chilled distilled water.

5. At the end of this time, remove the sample and measure the final volume.

6. Prepare the incubation mixtures in the following manner:

	Tube 1	Tube 2	Tube 3	Tube 4
Urine	0.8 ml	0.8 ml	0.8 ml	0.8 ml
p-Nitrophenyl-β-D-galactoside	0.2 ml	0.2 ml	- - -	- - -
p-Nitrophenyl-N acetyl-β-D-glucosaminide	- - -	- - -	0.2 ml	0.2 ml

7. Immediately place tubes #1 and #3 in an ice bath. These serve as blanks.

8. Place tubes #2 and #4 in a water bath maintained at 37°C.

9. At the end of 1 hour add 0.4 ml of 0.25M glycine buffer to each tube. Mix well.

10. Measure the absorption of each sample at 400 nm. Tube #1 serves as blank for tube #2 while #3 serves as a blank for #4.

Results:

1. The exact amount of p-nitrophenol released by the enzyme present in 0.8 ml of urine at 37°C. in 1 hour is obtained from the standard curve obtained by diluting samples of the p-nitrophenol standard stock solution.

2. Multiply the above answers by 1.25 to determine the amount of p-nitrophenol released by 1.0 ml of urine.

3. A correction should also be made for the dilution of the sample which occurs during dialysis.

4. The above calculations are made for both β-galactosidase and N-acetyl-β-glucosaminidase.

5. 1 unit of activity equals 1mμ mol of p-nitrophenol released by 1 ml of urine at 37°C. in 1 hour.

6. Relate the two enzyme levels as a ratio of N-acetyl-β-D-glucosaminidase to β-D-galactosidase.

Notes:

1. A well-characterized patient we have studied who had generalized gangliosidosis had an average β-galactosidase activity of 2.3 units (range = 0.7-3.5), while the control group had an average of 73.3 (range = 2.6-267.0). See also Table 14-1.

2. The ratio of the two enzymes in the patient was 356.8 (100-1,366), while the control group was 1.7 (0.2-19.8).

3. Although many other denominators could be conceived to which activity could be related (i.e. creatinine, nitrogen, grams solid, we have found volume alone appropriate for screening purposes.

TABLE 14-1.—β-Galactosidase, N-Acetyl-β-Glucosaminidase, and Ratio of N-Acetyl-β-Glucosaminidase to β-Galactosidase Activities in Urine Specimens From a Patient With Generalized Gangliosidosis, Patients With Mucopolysaccharidoses and a Control Group

	Number of Patients	Number of Specimens	Units of β-Galactosidase Activity*		Units of N-Acetyl-β-Glucosaminidase Activity*		Ratio of N-Acetyl-Glucosaminidase Galactosidase	
			Average	Range	Average	Range	Average†	Range
Generalized gangliosidosis (Patient J.Y.)	1	6	2.3	0.7 – 3.5	525.8	130.0 – 956.0	356.8	100.0 – 1,366.0
Mucopolysaccharidoses patients	7	9	16.4	8.0 – 49.3	92.1	35.0 – 178.0	6.2	0.4 – 15.0
Control patients	48	48	73.3	2.6 – 267.0	58.1	7.6 – 193.0	1.7	0.2 – 19.8

*1 unit equals 1 mμ mol of p-nitrophenol released by 1 ml of urine at 37°C. in 1 hour.
†These figures were determined from individual ratio values, not from the average enzyme levels.
From Thomas, G. H.: β-D-Galactosidase in human urine. Deficiency in generalized gangliosidosis, J. Lab. & Clin. Med. 74:725-731, 1969.

References

1. Sacrez, R., Juif, J. G., Gigonnet, J. M., and Gruner, J. E.: La maladie de Landing, Pédiatrie 22:143-162, 1967.

2. Okada, S., and O'Brien, J. S.: Generalized gangliosidosis. Beta-galactosidase deficiency, Science 160:1002-1004, 1968.

3. Sloan, H. R., Uhlendorf, B. W., Jacobson, C. B., and Fredrickson, D. S.: β-Galactosidase in tissue culture derived from human skin and bone marrow. Enzyme defect in GM_1 gangliosidosis, Pediat. Res. 3:532-537, 1969.

4. O'Brien, J.: Generalized gangliosidosis, J. Pediat. 75: 167-186, 1969.

5. Gatt, S.: Enzymatic hydrolysis of sphingolipids. V. Hydrolysis of monosialoganglioside and hexosylceramides by rat brain β-galactosidase, Biochem. Biophys. Acta 137:192-195, 1967.

6. Woollen, J. W., and Walker, P. G.: The fluorimetric estimation of N-acetyl-β-glucosaminidase and β-galactosidase in blood plasma, Clin. chim. acta 12:647-658, 1965.

7. Thomas, G. H.: β-D-Galactosidase in human urine. Deficiency in generalized gangliosidosis, J. Lab. & Clin. Med. 74:725-731, 1969.

CHAPTER 15

Hyperoxaluria: Urinary Oxalic Acid Procedure

Hyperoxaluria, the increased excretion of oxalic acid in urine, is found in at least two inherited disorders of metabolism. Each of these disorders is associated with calcium oxalate nephrolithiasis, chronic renal failure and, if untreated, death due to uremia (1). The clinical symptoms of these disorders appear to be due primarily to the formation and deposition of calcium oxalate crystals in several organs of the body, especially the kidneys (2).

Type I Hyperoxaluria

Evidence has been presented which indicates that the biochemical abnormalities in type I hyperoxaluria are due to a deficiency of the cytoplasmic 2-oxo-glutarate: glyoxylate carboligase (3). Associated with this enzyme defect is an excessive urinary excretion of oxalic, glycolic and glyoxylic acid.

Type II Hyperoxaluria

In contrast, type II hyperoxaluria is associated with an increased excretion of oxalic and glyceric acid while the excretion of glycolic acid is within the normal range. Moreover, the enzymatic defect in this disorder has been shown to be due to a dysfunction of D-glyceric dehydrogenase (4).

Method of Detection

As each of these disorders is associated with an increase in the urinary concentration of oxalic acid, it is possible to detect individuals with either of these disorders by simply measuring this compound in urine. A number of procedures have been described which should be suitable for this purpose (5-7).

We have found the procedure described by Archer et al. to be quite satisfactory for this purpose (5). In this method the oxalic acid present in the urine is precipitated as calcium oxalate by the addition of calcium chloride. The precipitated calcium oxalate is then washed, dissolved in 1N sulfuric acid and quantitated titrimetrically with potassium permanganate.

Urinary Oxalate

Method of Archer et al. (5)

Reagents:

1. Hydrochloric acid, concentrated.

2. Brom-cresol purple, 0.04% w/v aqueous solution: Dissolve 40 mg of this compound in a total of 100 ml of distilled water.

3. Ammonium hydroxide, 7.5N: Make a 1:1 dilution of concentrated NH_4OH with distilled water.

4. Acetic acid, 6N: Add 17.6 ml of glacial acetic acid to distilled water and dilute to a final volume of 50 ml.

5. pH Paper, narrow range, 4.0-5.5.

6. Calcium chloride, 5% w/v: Dissolve 10 Gm of $CaCl_2$ in water and dilute to a final volume of 200 ml.

7. Ammonium hydroxide, 0.35N: Add 2.3 ml of concentrated ammonium hydroxide to distilled water and dilute to a final volume of 100 ml.

8. Sulfuric acid, 1N: Add 5.4 ml of concentrated H_2SO_4 to distilled water and dilute to a final volume of 200 ml.

9. Potassium permanganate, 0.1N: Prepared standardized solutions may be obtained commercially, i.e., Fisher Scientific Company, Fairlawn, New Jersey.

10. Potassium permanganate, 0.01N: Dilute the stock 1:10 with distilled water. Place in an amber bottle and allow the solution to stand several days before standardizing.

11. Oxalic acid, 0.01N: Dry sodium oxalate in an oven at 100-105°C. for 12 hours and then allow to come to room temperature in a desiccator. Dissolve *exactly* 0.670 Gm. in distilled water in a 1-liter flask. Add 5 ml of concentrated sulfuric acid and dilute to a final volume of exactly 1 liter. Store in a refrigerator.

12. Potassium permanganate, standardized: Place exactly 5 ml of the 0.01N oxalic acid standard into a 50-ml Erlenmeyer flask and add 5 ml of 1N H_2SO_4. Heat to 70°C. in a water bath and titrate with 0.01N $KMnO_4$. Titrate a blank composed of 5 ml 1N H_2SO_4 and 5 ml distilled water, then subtract this result from the result obtained for the oxalic acid solution. Calculate and record the exact normality of the $KMnO_4$. Store the standardized solution in a refrigerator.

Procedure:

1. Obtain a 24-hour urine sample. This sample should be preserved with 5 ml of concentrated HCl per liter of urine.

2. Carefully measure the total volume of the urine, filter and place exactly 50 ml of the sample in a centrifuge tube with at least an 80-ml capacity.

3. Add 1 ml of Brom-cresol purple.

4. Add 7.5N ammonium hydroxide until the pH is alkaline (yellow to purple).

5. Add 6N acetic acid until the pH is 5.0-5.2 using pH paper.

6. Add 2 ml of the 5% calcium chloride solution.

7. Mix thoroughly and allow the solution to stand for about 16 hours at room temperature.

8. At the end of this time, centrifuge the solution for 20 minutes at 1,700 × g.

9. Transfer the precipitated calcium oxalate quantitatively to a 15-ml conical centrifuge tube using 0.35N ammonium hydroxide.

10. Centrifuge and carefully remove the ammonium hydroxide. Resuspend the precipitate again in the 0.35 N ammonium hydroxide.

11. Centrifuge as before, remove the NH_4OH solution and carefully allow the washed precipitate to drain.

12. Dissolve the precipitate in 2 ml of 1N sulfuric acid.

13. Warm the solution to 70°C. in a water bath.

14. Titrate with the standardized $KMnO_4$ from a 5-ml buret to a pink color which persists for at least 15 seconds.

15. Titrate in a similar manner a blank composed of 2.0 ml 1N H_2SO_4 and a standard composed of 1 ml of the 0.01N oxalic acid mixed with 2 ml of 1N H_2SO_4.

Calculations:

1. The calculation is based on the following:

   a. Formula weight of oxalic acid dihydrate (HOOC COOH · $2H_2O$) = 126.07.

   b. 1 ml of .01N $KMnO_4$ = 0.01 mEq.

   c. Oxalic acid has a valance of 2.

   d. $\dfrac{0.01 \text{ mEq} \times 126.07}{2}$ = 0.63035 mg oxalic acid dihydrate.

   e. 50-ml sample $\times$ 20 = 1 liter.

2. ___ ml $KMnO_4$ $\times$ 0.63035 mg $\times$ 20 = ___ mg oxalic acid dihydrate per liter of urine.

3. ___ mg oxalic acid per liter $\times$ volume in liters = ___ mg oxalic acid dihydrate per 24 hours concentration.

Notes:

1. Archer et al. reported the average excretion of oxalate dihydrate by normal adult male subjects to be 22 mg/24 hours.

2. Using this method we found a somewhat lower excretion (10-13 mg/24 hours) in a small number of children, while a patient with hyperoxaluria was found to excrete about ten times this amount (112-170 mg/24 hours).

3. Archer et al. have shown that the ingestion of increased amounts of oxalate will result in an increase in the urinary excretion of this compound. It is, therefore, recommended that foods rich in oxalate (rhubarb, strawberries, chocolate, cocoa, beets and spinach) be avoided prior to and during the testing period.

References

1. Williams, H. E., and Smith, L. H., Jr.: Primary Hyperoxaluria in Stanbury, J. B., Wyngaarden, J. B., and Fredrickson, D. S. (eds.): *The Metabolic Basis of Inherited Disease* (3rd ed.; New York: McGraw-Hill Book Company, Inc., 1972).

2. Williams, H. E., and Smith, L. H., Jr.: Disorders of oxalate metabolism, Am. J. Med. 45:715-735, 1968.

3. Koch, J., Stokstad, E. L. R., Williams, H. E., and Smith, L. H., Jr.: Deficiency of 2-oxo-glutarate: glyoxylate carboligase activity in primary hyperoxaluria, Proc. Nat. Acad. Sc. 57:1123-1129, 1967.

4. Williams, H. E., and Smith, L. H., Jr.: L-Glyceric aciduria. A new genetic variant of primary hyperoxaluria, New England J. Med. 278:233-239, 1968.

5. Archer, H. E., Dormer, A. E., Scowen, E. F., and Watts, R. W. E.: Studies on the urinary excretion of oxalate by normal subjects, Clin. Sc. 16:405-411, 1957.

6. Giterson, A. L., Slooff, P. A. M., and Schouten, H.: Oxalate in urine, Clin. chim. acta 29:342-343, 1970.

7. Hockaday, T. D. R., Frederick, E. W., Clayton, J. E., and Smith, L. H., Jr.: Studies on primary hyperoxaluria. II. Urinary oxalate, glycolate and glyoxylate measurement by isotope dilution methods, J. Lab. & Clin. Med. 65:677-687, 1965.

CHAPTER 16

Amino Acid Abnormalities: Medical and Laboratory Aspects

The detection, the identification and the quantitation of the amino acids in human body fluids have been carried out independently by a large number of workers. Serum and urine have been most extensively studied (1-7). More recently, however, spinal fluid (8), amniotic fluid (9), breast milk and cord plasma (10) have also been analyzed.

The information gained from these studies has resulted not only in the definition of normal values, but also in the recognition of a variety of physiological and pathological conditions associated with amino acid changes. These findings have been discussed in a number of reviews (11-24).

Clinical Importance

In the past 10 years mass screening of newborn infants and "high risk" groups for certain amino acid abnormalities has attracted much attention. The interest in these disorders results not from the magnitude of the problem in absolute numbers of affected persons, but rather from the consequences of the disorders. Specifically these include the following:

1. Severe retardation is a major risk of many inherited disorders of amino acid metabolism.

2. There is good evidence that the retardation in at least one of these disorders (phenylketonuria) can be prevented by dietary treatment if the disorder is identified in early infancy.

3. There are theoretical biochemical reasons to believe that the retardation may be prevented in other disorders by dietary and perhaps other therapeutic methods.

4. Much information regarding genetics and normal human biochemistry has been gained from these studies.

If one is to prevent the mental retardation, which appears to be a feature of a large number of these disorders, early diagnosis is essential. This requires that one detect the disorder in the "syndrome-free interval" just after birth, before irreversible damage is done.

While certain clinical findings of specific amino acid disorders may be suggestive, such as the maple syrup smell in branched-chain ketoaciduria, the liver cirrhosis in tyrosinemia and the ammonia intoxication of certain urea cycle defects, the majority of these disorders lack specific suggestive features. As a result the identification of affected patients is generally based on the laboratory demonstration of an abnormal aminoacidemia or aminoaciduria.

Amino acids are, of course, a normal constituent in serum and urine. The terms of aminoacidemia and aminoaciduria, therefore, have special meanings which are as follows:

1. Aminoacidemia is the presence in serum of one or more amino acids in quantities greater than normal or the presence of certain amino acids or intermediates of amino acids metabolism not usually found in serum.

2. Aminoaciduria is the urinary excretion of one or more amino acids in quantities greater than normal or the excretion of certain amino acids or intermediates of amino acid metabolism not usually found in the urine or both.

TABLE 16-1.–Inherited Disorders of Amino Acid Metabolism Associated With a Low Renal Clearance (High Threshold) of the Affected Amino Acid

Disorder	Amino Acid(s) Increased in Blood and Urine	References
Alaninemia	Alanine	25, 26
Argininemia	Arginine*	22, 27
Citrullinemia	Citrulline	28
Glycinemia, nonketotic	Glycine	29, 30
Glycinemia, ketotic†	Glycine	31, 32, 33
Histidinemia	Histidine	34
Hyperammonemia, type I	Glycine and glutamine	35
Hyperammonemia, type II	Glutamine	36
Hydroxyprolinemia	Hydroxyproline	37
Lysinemia, persistent	Lysine	38
Lysinemia, associated with hyperammonemia	Lysine and arginine	39
Maple syrup urine disease	Alloisoleucine, isoleucine, leucine and valine	40, 41
Methioinemia	Methionine	42
Oasthouse disease††	Isoleucine, leucine methionine, pheynlalanine and tyrosine	43
Ornithinemia	Ornithine	44
Phenylketonuria and its variants	Phenylalanine	45, 46
Pipecolatemia	Pipecolic acid	47
Prolinemia, type I	Proline §	48, 49
Prolinemia, type II	Proline §	50
Sarcosinemia	Sarcosine	51
Tyrosinemia	Tyrosine ¶	52
Tryptophanuria	Tryptophan	54
Valinemia	Valine	55

*Arginine, cystine, lysine and ornithine found in urine.

†May be due to a number of different specific defects including propionyl CoA carboxylase deficiency (32) and methylmalonic aciduria (33).

††It has been suggested that the Oasthouse disease may be identical to the methionine malabsorption syndrome (132). If this suggestion is confirmed this disorder may be more properly considered an inherited disorder of amino acid transport. (See Table 16-4.)

§ Proline, hydroxyproline and glycine increased in urine.

¶ Generalized increase in amino acids often found in urine (53).

Genetic Aminoacidemias

It is generally recognized that the aminoacidopathies in man can be divided into at least several categories. Within each of these categories are a number of disorders involving specific amino acid abnormalities. Gross amino acid abnormalities detected in properly stored serum are usually the result of an inherited defect. These amino acid disorders appear to result directly from a "metabolic block" which arises as a consequence of a genetically determined enzyme defect. In keeping with the one enzyme-one gene relationship, these abnormalities result from single gene defects and, in most cases, are inherited in an autosomal recessive manner. As there is a high renal threshold for many of the amino acids, one finds in many of these disorders a marked abnormality in both the serum and urine of affected patients. The inherited disorders of this type are shown in Table 16-1 (11,19,24).

Nongenetic Aminoacidemias

There is, in addition to these defects, a number of noninherited aminoacidemias (18). These changes are not, however, associated with a basic defect in amino acid metabolism, but rather, are related to transient changes in enzyme, cellular or organ function. Depending on the degree of the serum amino acid changes, there may also be a moderate to severe aminoaciduria. Examples of some of the conditions belonging to this group are listed in Table 16-2.

Low-Threshold Aminoacidurias

In addition to the aminoacidemias, there is also a large group of amino acid abnormalities which are most prominent in the urine (19,24). Among these are the so-called "low-threshold aminoacidurias" (Table 16-3). These disorders also result from the block of normal metabolic pathways due to inherited enzyme defects. The major difference between these disorders and the inherited primary aminoacidemias is that here the amino acid, whose metabolism is blocked, does not rise to a high concentration in the serum due to a lack of active reabsorption by the kidney. Thus, in contrast to the usual aminoacidemias, the excessive amino acids are present primarily in the urine while the serum concentrations are essentially normal. It is important to

TABLE 16-2.—Noninherited Conditions Associated With Changes in Serum Amino Acids

Condition	Amino Acid Changes	References
Transient tyrosinemia of premature infants	Tyrosine	56
Neonatal hypermethioninemia	Hyperaminoacidemia, methionine most prominent	57
Neonatal hyperhydroxyprolinemia	Hydroxyproline	58
Neonatal hyperphenylalaninemia	Phenylalanine*	11
Liver necrosis	Hyperaminoacidemia, tyrosine & methionine most prominent	18
Certain infections	Varies with disease and stage of disorder	18

*Often occurs with tyrosinemia (11).

TABLE 16-3.—Inherited Disorders of Amino Acid Metabolism Associated With
a High Renal Clearance (Low Threshold) of the Affected Amino Acid

Disorder	Amino Acids Increased in Urine	References
β-Alaninemia	β-Aminoisobutyric acid, γ-aminobutyric acid and taurine	59
β-Aminoisobutyric aciduria	β-Aminoisobutyric acid	60
Argininosuccinic aciduria	Argininosuccinic acid & citrulline	61
Aspartylglycosaminuria	Aspartylglycosamine	62
Carnosinemia	Carnosine	63
Cystathionuria	Cystathionine	64
Hypophosphatasia	Phosphoethanolamine	65
Homocystinuria	Homocystine	66
Imidazole amino aciduria	Carnosine, anserine, homocarnosine & methylhistidine	67
β-mercaptolactate-cysteine disulfiduria	Mercaptolactate-cysteine disulfide	68, 69
Sulfite oxidase deficiency	Sulfocysteine	70

to understand that the lack of active kidney transport is not due to a kidney defect, but is rather the normal situation with regard to the altered compounds.

Inherited Primary Aminoacidurias

There is also a second group of inherited aminoacidurias which results, not from a metabolic block, but rather from a reduction in the normal capacity of the kidney to reabsorb certain amino acids (12,21). The evidence for a defect in the active transport of these amino acids is rather strong; however, the actual specific mechanism at the cellular level is not yet clearly understood.

Recent evidence indicates that the transport of amino acids between the urine and serum involves a variety of different active systems. Evidence for these various pathways include concentration against a gradient, chemical and optical specificity, and energy and ion dependence (17,71). These systems thus have a certain stereochemical specificity for the various amino acids. As a consequence, one often finds that the disruption of one system results in an alteration in the urinary excretion of several chemically related amino acids (Table 16-4).

Inherited Secondary Aminoacidurias

The disorders listed in Table 16-4 are the result of genetic defects directly affecting transport systems in the kidney and other organs. There are, in addition, a number of aminoacidurias which result from the secondary deleterious effects of inherited disorders not directly related to renal transport systems (18,19,24,53). These disorders do, however, have at least one thing in common, namely, a major involvement of the kidney as part of the disease process. One thus finds in the presence of normal serum amino acids, a generalized aminoaciduria which results from secondary tubular abnormalities (Table 16-5).

TABLE 16-4.—Inherited Disorders of Amino Acid Transport

Disorder	Amino Acids Increased in Urine	References
Cystinuria, types I, II, III	Arginine, cystine, lysine and ornithine	72
Dibasic aminoaciduria	Arginine, lysine and ornithine	73
Hartnup disease	Neutral aminoaciduria (except imino acids and glycine)	74
Hypercystinuria	Cystine	21, 75
Iminoglycinuria	Glycine, hydroxyproline and proline	76
Luder-Sheldon syndrome	Generalized aminoaciduria	77
Methionine malabsorption	Methionine, smaller amounts of tyrosine, phenylalanine, valine and leucine	78
Glucoglycinuria	Glycine	79
Glycinuria with hypophosphatemia	Glycine	80
Glycinuria*	Glycine	81
Familial protein intolerance	Lysine, also sometimes arginine	83

*May reflect the heterozygous phenotype of the iminoglycinuric disorder (82).

The exact mechanism involved in these disorders is poorly understood at the present time. In some, for example galactosemia and fructose intolerance, there is evidence that the accumulation of sugar phosphates in the tubules may cause the aminoaciduria, while in Wilson's disease, the aminoaciduria may result from cellular damage arising from the cellular deposition of large amounts of copper. In any event, the removal of these various compounds frequently cures the aminoaciduria.

TABLE 16-5.—Inherited Disorders Associated With Secondary Aminoacidurias Presumably Due to Progressive Impairment of Renal Tubules

Disorder	Amino Acids Increased in Urine	References
"Busby" syndrome	Generalized aminoaciduria	21, 84, 85
Cystinosis	Generalized aminoaciduria	53, 86
Fanconi syndrome (idiopathic)	Generalized aminoaciduria	53, 87
Fructose intolerance	Generalized aminoaciduria	53
Galactosemia	Generalized aminoaciduria	88
Glycogen storage disease, type I (rarely)	Generalized aminoaciduria	89
Lactose intolerance	Generalized aminoaciduria	90
Lowe's syndrome	Generalized aminoaciduria	91, 92
Tyrosinosis	Generalized aminoaciduria	53, 52
Wilson's disease	Generalized aminoaciduria	93, 94

TABLE 16-6.—Noninherited, Transient and/or Acquired Aminoacidurias

Condition	Amino Acid(s) Increased in Urine	References
Aminoaciduria of normal newborns	Generalized	95,96
Body irradiation	Generalized	97
Connective tissue disorders	Hydroxyproline	19
Heavy metal toxicity	Generalized	98
Hepatic necrosis	Generalized, especially tyrosine, leucine, & proline	18
Homocitrullinuria	Homocitrulline	99
Hyperparathyroidism	Hydroxyproline	100
Hyperthyroidism	Hydroxyproline	101
Infectious disorders (?)	Generalized	102
Malnutrition (?)	Generalized	103
Neurosecretory tumors	Cystathionine	104
Nephrotic syndrome	Generalized	105
Outdated tetracycline ingestion	Generalized	106
Pregnancy	Generalized	18
Renal transplantation reaction	Generalized	18
Salicylate toxicity	Generalized	107
Steroid therapy	Generalized	
Thermal burn	Generalized	108
Vitamin B deficiency	Cystathionine	24
Vitamin C deficiency	Generalized	109
Vitamin D deficiency	Generalized	110

Noninherited Secondary Aminoacidurias

There are also, in addition, a rather large number of secondary aminoacidurias which are not inherited (18,24). Here again the amino acid abnormality appears to reside in the kidney tubules, thus resulting in normal serum amino acid concentrations and generalized aminoacidurias. The abnormalities apparently have, as their basis, cellular or organ alterations resulting from noninherited factors. These range from immaturity of the transport system through thermal body burns and irradiation to the toxic effects of a variety of compounds (Table 16-6).

Maternal PKU

Another complex relationship between an amino acid abnormality and mental retardation has also been demonstrated in maternal PKU. In this situation intrauterine growth retardation, neurological damage and congenital abnormalities have been found in the non-phenylketonuric offspring of phenylketonuric mothers (111,112). Current evidence strongly suggests that this damage is the consequence, directly or indirectly, of the high maternal blood levels of phenylalanine or its metabolites during gestation (9).

Ninhydrin Reaction

Almost all the techniques for the detection and quantitation of the various amino acids utilize the chemical compound ninhydrin. This compound reacts with the a-amino nitrogen group to yield a colored reaction product which has a maximum absorption at either 440 or 570 nm. As this is a general reaction, however, it requires that one employ some technique to separate all ninhydrin positive compounds from each other in order to obtain a quantitative estimate of each individual amino acid. One such method is the semiquantitative technique of paper chromatography of amino acids developed by Dent (113). This method still remains in wide use in many modified forms throughout the world at the present time (114-117).

Column Chromatography

Another widely used technique is that of column chromatography first developed by Moore and Stein (118). These columns, in combination with automatic recording devices, have made it possible to routinely measure quantitatively the amino acids in the biological fluids (119). More recent improvements in ion exchange resins, buffers and high sensitivity detection systems have resulted in a number of improved column chromatography systems for the rapid separation of the amino acids (6,8,120-124).

In addition, a number of other methods have also been employed to study amino acids in biological fluids. These include high-voltage electrophoresis (19,125-128), thin layer chromatography (24) and gas chromatography (129).

Normal Findings in Serum

While Hamilton has demonstrated that there may be as many as 50 ninhydrin positive compounds in serum, most of the above methods allow one to separate routinely and identify about 20-25 compounds (122). The majority of the amino acids are present in concentrations of about 1-3 mg/100 ml of fluid. Glutamine, however, is present at about 8 mg/100 ml in fresh serum samples. In addition, a small number of compounds are usually found in trace amounts.

The range of normal values in serum for any given amino acid is generally quite small, being in most cases no more than about ± 2 mg/100 ml. It should also be noted that while differences do exist between newborns and adults, these differences are generally small. Exceptions to this are given in Table 16-2. There is little, if any, difference between the serum amino

---

TABLE 16-7.—Alterations in Amino Acid Concentrations Which May Be Introduced by
Laboratory Procedures

I.   Artifacts which may arise during sample preparation.

    a.   Binding of cystine, and/or homocystine by plasma proteins due to delay in deproteinization of sample (1).

    b.   Increased concentrations of aspartic acid, glutamic acid, ornithine, phosphoethanolamine and taurine due to hemolysis of sample (130).

    c.   Decreased concentrations of cystine and ornithine due to hemolysis (130).

    d.   Increased concentrations of aspartic acid, glutamic acid, phosphoethanolamine and taurine due to contamination of the sample with platelets and leukocytes (130).

    e.   Changes in the concentrations of various amino acids due to the contamination with hands and fingers (131).

II.   Artifacts which may arise during sample storage at -20°C.

    a.   Increased concentrations of aspartic acid, glutamic acid and methionine sulfoxide (3).

    b.   Decreased concentrations of asparagine, glutamine, methionine and tryptophan (3).

III.   Artifacts which may arise during analysis of sample.

    a.   Conversion of glutamine to pyrrolidone carboxylic acid during column chromatography (3).

    b.   Conversion of arginosuccinic acid to 5 and 6-membered ring anhydrides.

---

acid concentrations in males and females. There do appear, however, to be small but consistent diurnal variations in serum amino acid levels. The highest concentrations are seen daily during the active period and the lowest during sleep (18). Small changes with menstrual cycle and with changes in the diet have also been reported (18).

While these normal variations may offer much information in regard to the study of amino acid regulation and metabolism, it is also true that they are sufficiently small so as to present no great problem in the study of known genetic diseases.

It should also be emphasized here that quantitative changes will also occur in serum, as well as in other biological fluids, if the sample is not handled or stored appropriately (3,6,130). Some of these changes have been discussed by a number of investigators and their findings are listed in Table 16-7.

Normal Findings in Urine

In contrast to the situation with serum, there are in urine, by most systems of chromatography, a number of ninhydrin-positive compounds which may be difficult to identify. Perhaps even more troublesome is the fact that detailed analysis has indicated that there may be as many as 130 compounds in urine (122), while most standard methods give evidence for only 50-60. This suggests that many peaks which have been thought to represent a single compound may, in many instances, represent several poorly resolved ninhydrin positive compounds. Yet another problem is that of urine volume changes, which is often corrected for by relating the concentration of the amino acids to the creatinine levels or to total urine solids.

Paper Chromatography Screening Methods

Most laboratories require a rapid, simple chromatography method which will allow one to screen a large population for amino acid abnormalities. For this we recommend the single dimension paper method. A major virtue of this method is its simplicity. A large number of samples can be easily run by a single technician with only inexpensive equipment. The separation of individual amino acids and the ability to quantitate them is, of course, considerably poorer than with the column ion exchange method. By and large, however, most known amino acid abnormalities are associated with large changes in amino acid concentrations and are, therefore, readily detectable by this technique.

For applications requiring specific information, we recommend an ion exchange method using one of the improved automated systems.

One Dimension Chromatographic Procedure for Amino Acids in Serum

Method modified from Efron et al. (115) and Scriver et al. (116)

Reagents and Equipment:

1. Chromatography medium: Whatman 3MM paper, 46 × 57 cm. Narrower paper can be used, depending on the number of samples to be run.

2. Chromatography tank: A large model which will accommodate up to 10 sheets of paper has been found to be suitable.

3. Paper punch: Commercial type for notebook paper with a 1/4″ diameter.

4. Staining trays.

5. Chromatography solvent: Prepare a mixture of butanol-acetic acid-water in a ratio of 12:3:5.

6. Ninhydrin/isatin stain: Prepare by adding 2.5 Gm of ninhydrin, 0.1 Gm of isatin and 10 ml of lutidine to acetone. Dilute to a final volume of 1,000 ml with acetone. Store in a dark bottle in a refrigerator. Prepare fresh weekly.

7. Ehrlich's reagent: Dissolve 10 Gm of p-dimethylaminobenzaldehyde in concentrated HCl. Dilute to a final volume of 100 ml with HCl. Add 400 ml of acetone just before use.

8. Pauly's reagent: Prepare this stain as follows:

    a. Sulphanilic acid: Dissolve 9 Gm of sulphanilic acid in a mixture of 90 ml of concentrated HCl and 900 ml of distilled water.

    b. Sodium nitrite, 5%: Dissolve 5 Gm of sodium nitrite in water and dilute to a final volume of 100 ml with water.

    c. Sodium carbonate, 10%: Dissolve 20 Gm of anhydrous sodium carbonate in water and dilute to a final volume of 200 ml with water.

TABLE 16-8.—Preparation of the Amino Acid Standards for the One-Dimension Chromatographic Procedure

Amino Acid	Molecular Weight	For $1\,\mu m/ml$ Standards mg of Amino Acid to Mix With 1 ml H_2O to Make			For Standards With Normal Plasma Concentrations	
		40 $\mu m/ml$	20 $\mu m/ml$	10 $\mu m/ml$	Normal Concentration Plasma AAs $\mu m/ml$	mg for 10 ml H_2O to Give This Concentration
Alanine	89.1	3.56			0.4	14.25
Arginine	210.7	8.42			0.1	8.42
Aspartic	133.1		2.66		0.01	0.27
Cystine	240.3	9.61			0.05	4.8
Glycine	75.1	3.00			0.25	7.50
Glutamic	147.1		2.94		0.05	1.47
Histidine	209.6	8.38			0.075	6.28
Isoleucine	131.1	5.24			0.06	3.14
Leucine	131.2	5.24			0.125	6.55
Lysine	182.7	7.30			0.15	10.95
Methionine	149.2	5.96			0.025	1.49
Proline	115.1	4.60			0.2	9.2
Phenylalanine	165.2	6.60			0.05	3.30
Serine	105	4.20			0.13	5.46
Threonine	119	4.76			0.16	7.62
Tryptophan	204	8.16				
Tyrosine	181			1.812	0.05	0.91
Valine	117	4.68			0.2	9.36
Asp - NH_2	132	5.28				
Glu - NH_2	146	5.84				

To make a solution containing 1 $\mu m/ml$ of the above amino acids:
    Add 0.5 ml of the 40 $\mu m/ml$ solutions of amino acids, 1.0 ml of the aspartic and glutamic (20-$\mu m/ml$ solutions) and 2.0 ml of the 10 $\mu m/ml$ solution of tyrosine; then make to a final volume of 20 ml with water.

Normal Plasma Concentration Standard:
    Add together 0.5 ml of the solutions made in the concentrations of the last column with the exception of aspartic, glutamic and tyrosine, 1.0 ml of aspartic and glutamic, and 2.0 ml of the tyrosine solution. Make up to total volume of 20 ml.

    d. Prepare the complete stain, just before needed, by mixing together 100 ml of each of the first two solutions (a & b). Allow this mixture to stand at room temperature for 4-5 minutes. At this time carefully add 200 ml of the 10% sodium carbonate to the mixture.

9. Amino acid standards: See Table 16-8.

Method:

1. Spot serum samples on 2″ × 3″ pieces of 3MM paper and allow to air dry.

2. Punch holes 2.5 cm apart, 2 cm from the bottom of the chromatograph paper sheets.

3. Punch discs from the papers containing the dried serum samples and amino acid standards using the same punch. Unlike the whole blood method used by Efron et al. (115), no autoclaving of the samples is necessary.

4. Carefully insert these discs in the previously punched holes in the chromatograph sheets. These are pressed in by running a smooth glass stirring rod around the outer margins. Discs stay in place during the chromatographic run without problems.

5. Use disposable plastic gloves when handling the chromatogram and serum discs to prevent contamination by ninhydrin positive material from the hands.

6. Develop the chromatogram in an ascending manner using the butanol-acetic acid-water solvent at room temperature for approximately 19 hours.

7. At the end of this time remove the chromatogram sheets and air dry in a flume hood.

Fig. 16-1.—One dimension chromatographic procedure for amino acids in serum.

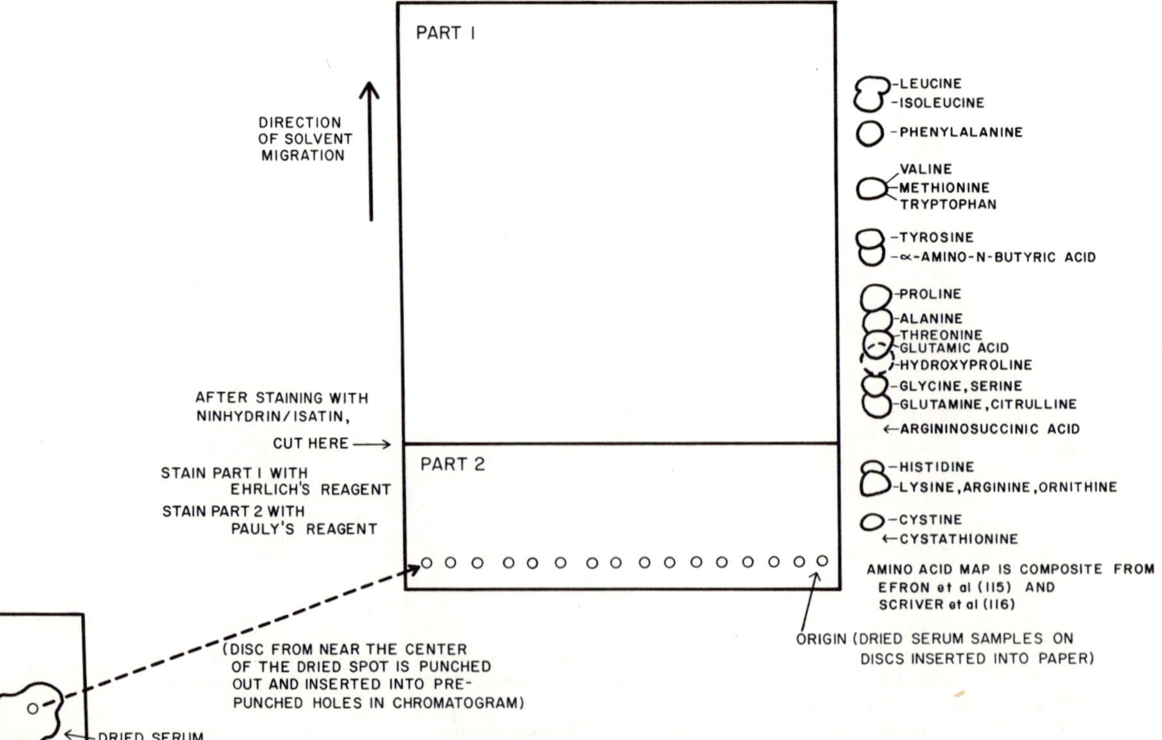

8. Dip the sheets through the ninhydrin/isatin stain using the staining trays. Dry the chromatograms in an oven at 80°C. for 10 minutes.

9. Observe the chromatogram for the presence of abnormal patterns using the normal control serums and the amino acid standards as markers. See Figure 16-1.

10. Chromatograms can be overstained with Pauly's reagent to show histidine and Ehrlich's reagent to identify hydroxyproline and citrulline in the following manner:

   a. Carefully reheat the chromatogram at 80°C. for 10 minutes to remove any phenol.

   b. Cut the paper as indicated in Figure 16-1.

   c. Dip the upper portion (1) through Ehrlich's stain and the lower part (2) through Pauly's stain.

   d. The exact position to cut the chromatogram should be established for each laboratory by locating histidine on a standard and always cutting the papers so this will be in the lower part as shown in Figure 16-1.

Notes:

1. The results of this procedure are shown in Table 16-9.

TABLE 16-9.—Color Reactions of the Various Amino Acids Found in Serum*

Amino Acids†	Ninhydrin-Isatin	Other Reagents
Isoleucine	Purple	
Leucine		
Phenylalanine	Grey-brown	
Valine	Purple	
Methionine		
Tryptophan		
Tyrosine	Brown-grey	
a-Amino-n-butyric acid	Blue-purple	
Proline	Yellow—darkens with standing	
Alanine	Purple	
Threonine	Purple	
Glutamic acid		
Hydroxyproline	Magenta	Red-purple (Ehrlich's)
Glycine	Brown-orange	
Serine	Purple-grey	
Glutamine	Purple	Brown-orange (Ehrlich's)
Citrulline	Purple	
Aspartic acid	Blue	
Argininosuccinic acid	Grey-purple	
Histidine	Grey	Brown-red (Pauly)
Arginine	Purple	
Lysine		
Ornithine		
Cystathionine	Grey-blue	

*From Scriver, C. R., Davies, E., and Cullen, A. M.: Application of a simple micromethod to the screening of plasma for a variety of aminoacidopathies, Lancet 2:230–232, 1964.

†Grouped according to chromatographic separation.

2. This is a screening test only, and any suspected abnormalities should be verified by more accurate techniques such as ion exchange chromatography.

References

1. Stein, W. H., and Moore, S.: The free amino acids of human blood plasma, J. Biol. Chem. 211:915-926, 1954.

2. Armstrong, M.D., Yates, K. N., and Connelly, J. P.: Amino acid excretion of newborn infants during the first twenty-four hours of life, Pediatrics 33:975-978, 1964.

3. Dickinson, J. C., Rosenblum, H., and Hamilton, P. B.: Ion exchange chromatography of the free amino acids in the plasma of the newborn infant, Pediatrics 36:2-13, 1965.

4. Brodehl, J., Gellissen, K., and Jäkel, A.: Endogenous renal transport of free amino acids in infancy and childhood, Pediatrics 42:395-404, 1968.

5. Peters, J. H., Lin, S. C., Berridge, B. J., Jr., Cummings, J. G., and Chao, W. R.: Amino acids, including asparagine and glutamine, in plasma and urine of normal human subjects, Proc. Soc. Exper. Biol. & Med. 131: 281-288, 1969.

6. Peters, J. H., and Berridge, B. J., Jr.: The determination of amino acids in plasma and urine by ion-exchange chromatography, Chromatog. Rev. 12:157-165, 1970.

7. Dickinson, J. C., Rosenblum, H., and Hamilton, P. B.: Ion exchange chromatography of the free amino acids in the plasma of infants under 2,500 Gm at birth, Pediatrics 45:606-613, 1970.

8. Dickinson, J. C., and Hamilton, P. B.: The free amino acids of human spinal fluid determined by ion exchange chromatography, J. Neurochem. 13:1179-1187, 1966.

9. Thomas, G. H., Parmley, T. H., Stevenson, R. E., and Howell, R. R.: Developmental changes in amino acid concentrations in human amniotic fluid. Abnormal findings in maternal phenylketonuria, Am. J. Obst. & Gynec. 111:38-42, 1971.

10. Ghadimi, H., and Pecora, P.: Free amino acids of cord plasma as compared with maternal plasma during pregnancy, Pediatrics 33:500-511, 1964.

11. Scriver, C. R., Clow, C. L., and Lamm, P.: Plasma amino acids. Screening, quantitation and interpretation, Am. J. Clin. Nutr. 24:876-890, 1971.

12. Hollerman, C. E., and Calcagno, P. L.: Aminoaciduria-renal transport, Am. J. Dis. Child. 115:169-178, 1968.

13. Efron, M. L.: Aminoaciduria, New England J. Med. 272:1058-1066 & 1107-1113, 1965.

14. Nyhan, W. L., and Tocci, P.: Aminoaciduria, Ann. Rev. Med. 17:133-160, 1966.

15. Efron, M. L., and Ampola, M. G.: The aminoacidurias, Pediat. Clin. North America 14:881-903, 1967.

16. Ghadimi, H.: Diagnosis of inborn errors of amino acid metabolism, Am. J. Dis. Child. 114:433-439, 1967.

17. Scriver, C. R.: Inborn errors of amino-acid metabolism, Brit. M. Bull. 25:35-41, 1969.

18. Feigin, R. D.: Blood and urine amino acid aberrations. Physiologic and pathological changes in patients without inborn errors of amino acid metabolism, Am. J. Dis. Child. 117:24-47, 1969.

19. Mabry, C. C.: High voltage electrophoresis, CRC Critical Reviews, Clinical Laboratory Sciences 1:135-190, 1970.

20. O'Brien, D.: *Rare Inborn Errors of Metabolism in Children with Mental Retardation*, PHS Publication No. 2049, 1970.

21. Scriver, C. R., and Hechtman, P.: Human Genetics of Membrane Transport With Emphasis on Amino Acids in Harris, H., and Hirschhorn, K. (eds.): *Advances in Human Genetics* (Vol. I; New York: Plenum Press, 1970).

22. Levin, B.: Hereditary metabolic disorders of the urea cycle, Advances Clin. Chem. 14:65-143, 1971.

23. Menkes, J. H.: Disorders of amino acid metabolism, California Med. 115:14-23, 1971.

24. Saifer, A.: Rapid screening methods for the detection of inherited and acquired aminoacidopathies, Advances Clin. Chem. 14:145-218, 1971.

25. Tada, K., Yoshida, T., Konno, T., Wada, Y., Yokoyama, Y., and Arakawa, T.: Hyperalaninemia with pyruvicemia, Tohoku. J. Exper. Med. 97:99-100, 1969.

26. Lonsdale, D., Faulkner, W. R., Price, J. W., and Smeby, R. R.: Intermittent cerebellar ataxia associated with hyperpyruvic acidemia, hyperalaninemia, and hyperalaninuria, Pediatrics 43:1025-1034, 1969.

27. Terheggen, H. G., Schwenk, A., Lowenthal, A., Van Sande, M., and Colombo, J. P.: Hyperargininamie mit Arginasedefekt. Eine neue familiare Stoffwechselstorung, Ztschr. Kinderh. 107:298-312 & 313-323, 1970.

28. McMurray, W. C., Mohyuddin, F., Rossiter, R. J., Rathbun, J. C., Valentine, G. H., Koegler, S. J., and Zarfas, D. E.: Citrullinuria. A new aminoaciduria associated with mental retardation, Lancet 1:138, 1962.

29. Gerritsen, T., Kaveggia, E., and Waisman, H. A.: A new type of idiopathic hyperglycinemia with hypoxaluria, Pediatrics 36:882-891, 1965.

30. Nyhan, W. L.: Nonketotic Hyperglycinemia in Stanbury, J. B., Wyngaarden, J. B., and Fredrickson, D. S. (eds.): *The Metabolic Basis of Inherited Disease* (3rd ed.; New York: McGraw-Hill Book Company, Inc., 1972).

31. Childs, B., Nyhan, W. L., Borden, M., Bard, L., and Cooke, R. E.: Idiopathic hyperglycinemia and hyperglycinuria. A new disorder of amino acid metabolism. I., Pediatrics 27:522-538, 1961.

32. Hsia, Y. E., Scully, K. J., and Rosenberg, L. E.: Defective propionate carboxylation in ketotic hyperglycinemia, Lancet 1:757-758, 1969.

33. Oberholzer, V. G., Levin, B., Burgess, E. A., and Young, W. F.: Methylmalonic acidosis. An inborn error of metabolism leading to chronic metabolic acidosis, Arch. Dis. Childhood 42:492-504, 1967.

34. Ghadimi, H., Partington, M. W., and Hunter, A.: A familial disturbance of histidine metabolism, New England J. Med. 265:221-224, 1961.

35. Freeman, J. M., Nicholson, J. F., Masland, W. S., Rowland, L. P., and Carter, S.: Ammonia intoxication due to a congenital defect in urea synthesis, J. Pediat. 65:1039-1040, 1964.

36. Russell, A., Levin, B., Oberholzer, V. G., and Sinclair, L.: Hyperammonemia, a new instance of an inborn enzymatic defect of the biosynthesis of urea, Lancet 2:699-700, 1962.

37. Efron, M. L., Bixby, E. M., Palattao, L. G., and Pryles, C. V.: Hydroxyprolinemia associated with mental deficiency, New England J. Med. 267:1193-1194, 1962.

38. Ghadimi, H., Binnington, V. I., and Pecora, P.: Hyperlysinemia associated with mental retardation, New England J. Med. 273:723-728, 1965.

39. Colombo, J. C., Vassella, F., Humbel, R., and Buergi, W.: Lysine intolerance with periodic ammonia intoxication, Am. J. Dis. Child. 113:138-141, 1967.

40. Menkes, J. H., Hurst, P. L., and Craig, J. M.: A new syndrome. Progressive familial infantile cerebral dysfunction associated with an unusual urinary substance, Pediatrics 14:462-467, 1954.

41. Morris, M. D., Lewis, B. D., Doolan, P. D., and Harper, H. A.: Clinical and biochemical observations on an apparently nonfatal variant of branched chain ketoaciduria (maple syrup urine disease), Pediatrics 28:918-923, 1961.

42. Perry, T. L., Hardwick, D. F., Dixon, G. H., Dolman, C. L., and Hansen, S.: Hypermethioninemia. A metabolic disorder associated with cirrhosis, islet cell hyperplasia, and renal tubular degeneration, Pediatrics 36:236-250, 1965.

43. Smith, A. J., and Strang, L. B.: An inborn error of metabolism with the urinary excretion of α-hydroxybutyric acid and phenylpyruvic acid, Arch. Dis. Childhood 33:109-113, 1958.

44. Shih, V. E., Efron, M. L., and Moser, H. W.: Hyperornithinemia, hyperammonemia, and homocitrullinuria. A new disorder of amino acid metabolism associated with myoclonic seizures and mental retardation, Am. J. Dis. Child. 117:83-92, 1969.

45. Folling, A.: Über Ausscheidung von Phenylbrenztraubensäure in den Harn als Stoffwechselanomalie in Verbindung mit Imbezillität, Hoppe-Seyler's Zeitschrift für Physiologische Chemie 227:169-176, 1934.

46. Auerbach, V. H., DiGeorge, A. M., Carpenter, G. G., and Wood, P.: Phenylalaninemia in Nyhan, W. L. (ed.): *Amino Acid Metabolism and Genetic Variation* (New York: McGraw-Hill Book Company, 1967).

47. Gatfield, P. D., Taller, E., Hinton, G. G., Wallace, A. C., Abdelnour, G. M., and Haust, M. D.: Hyperpipecolatemia. A new metabolic disorder associated with neuropathy and hepatomegaly. A case study. Canad. M. A. J. 99:1215-1233, 1968.

48. Shafer, L. A., Scriver, C. R., and Efron, M. L.: Familial hyperprolinemia, cerebral dysfunction and renal anomalies occurring in a family with hereditary nephropathy and deafness, New England J. Med. 267:51-60, 1962.

49. Woody, N. C., Snyder, C. H., and Harris, J. A.: Hyperprolinemia. Clinical and biochemical family study, Pediatrics 44:554-563, 1969.

50. Emery, F. A., Goldie, L., and Stern, J.: Hyperprolinaemia type 2, J. Ment. Defic. Res., 12:187-195, 1968.

51. Gerritsen, T., and Waisman, H. A.: Hypersarcosinemia. An inborn error of metabolism, New England J. Med. 275:66-69, 1966.

52. Gentz, J., Jagenburg, R., and Zetterström, R.: Tyrosinemia, J. Pediat. 66:670-696, 1965.

53. Schneider, J. A., and Seegmiller, J. E.: Cystinosis and the Fanconi Syndrome, in Stanbury, J. B., Wyngaarden, J. B., and Fredrickson, D. S. (eds.): *The Metabolic Basis of Inherited Disease* (3rd ed.; New York: McGraw-Hill Book Company, Inc., 1972).

54. Tada, K., Ito, H., Wada, Y., and Arakawa, T.: Congenital tryptophanuria with dwarfism, Tohoku J. Exper. Med. 80:118-133, 1963.

55. Wada, Y., Tada, K., Minagawa, A., Yoshida, T., Morikawa, T., and Okamura, T.: Idiopathic hypervalinemia probably a new entity of inborn error of valine metabolism, Tohoku J. Exper. Med. 81:46-55, 1963.

56. Menkes, J. H., and Avery, M. E.: The metabolism of phenylalanine and tyrosine in the premature infant, Bull. Johns Hopkins Hosp. 113:301-319, 1963.

57. Levy, H. L., Shih, V. E., Madigan, P. M., Karolkewicz, V., Carr, J. R., Lum, A., Richards, A. A., Crawford, J. D., and MacCready, R. A.: Hypermethioninemia with other hyperaminoacidemias. Studies in infants on high protein diets, Am. J. Dis. Child. 117:96-103, 1969.

58. Marrow, G., Kivirikko, K. I., and Prockop, D. J.: Hydroxyprolinemia and increased excretion of free hydroxyproline in early infancy, J. Clin. Endocrinol. 26:1012-1014, 1966.

59. Scriver, C. R., Pueschel, S., Davies, E.: Hyper-β-alaninemia associated with β-aminoaciduria and γ-aminobutyricaciduria, somnolence and seizures, New England J. Med. 274:635-643, 1966.

60. Armstrong, M. D., Yates, K., Kakimoto, Y., Taniguchi, K., and Kappe, T.: Excretion of β-amino-isobutyric acid by man, J. Biol. Chem. 238:1447-1455, 1963.

61. Westall, R. G.: Argininosuccinic aciduria. Identification and reactions of the abnormal metabolite in a newly described form of mental disease, with some preliminary metabolic studies, Biochem. J. 77:135-144, 1960.

62. Pollitt, R. J., Jenner, F. A., and Merskey, H.: Aspartylglycosaminuria. An inborn error of metabolism associated with mental defect, Lancet 2:253-255, 1968.

63. Perry, T. L., Hansen, S., Tischler, B., Bunting, R., and Berry, K.: Carnosinemia. A new metabolic disorder associated with neurologic disease and mental defect, New England J. Med. 277:1219-1227, 1967.

64. Harris, H., Penrose, L. S., and Thomas, D. H. H.: Cystathioninuria, Ann. Human Genetics 23:442-453, 1958-59.

65. McCance, R. A., Morrison, A. B., and Dent, C. E.: Excretion of phosphoethanolamine and hypophosphatasia. Preliminary communication, Lancet 1:131, 1955.

66. Carson, N. A., Cusworth, D. C., Dent, C. E., Field, C. M. B., Neill, D. W., and Westall, R. G.: Homocystinuria. A new inborn error of metabolism associated with mental deficiency, Arch. Dis. Childhood 38:425-436, 1963.

67. Bessman, S. P., and Baldwin, R.: Imidazole aminoaciduria in cerebromacular degeneration, Science 135: 789-791, 1962.

68. Ampola, M. G., Efron, M. L., Bixby, E. M., and Meshorer, E.: Mental deficiency and a new aminoaciduria, Am. J. Dis. Child. 117:66-70, 1969.

69. Crawhall, J. C., Parker, R., Sneddon, W., and Young, E. P.: Beta-mercaptolactate-cysteine disulfide in the urine of a mentally retarded patient, Am. J. Dis. Child. 117: 71-82, 1969.

70. Irreverre, F., Mudd, S. H., Heizer, W. D., and Laster, L.: Sulfite oxidase deficiency. Studies of a patient with mental retardation, dislocated ocular lenses, and abnormal urinary excretion of S-sulfo-L-cysteine, sulfite, and thiosulfate, Biochem. Med. 1:187-217, 1967.

71. Scriver, C. R.: Amino acid transport. Factors influencing uptake in kidney and intestine, Mod. Prob. in Pediat. 11:72-91, 1968.

72. Thier, S. O., and Segal, S.: Cystinuria in Stanbury, J. B., Wyngaarden, J. B., and Fredrickson, D. S. (eds.): *The Metabolic Basis of Inherited Disease* (3rd ed.; New York: McGraw-Hill Book Company, Inc., 1972).

73. Whelan, D. T., and Scriver, C. R.: Hyperdibasicaminoaciduria. An inherited disorder of amino acid transport, Pediat. Res. 2:525-534, 1968.

74. Scriver, C. R.: Hartnup disease. A genetic modification of intestinal and renal transport of certain neutral alpha-amino acids, New England J. Med. 273:530-532, 1965.

75. Brodehl, J., Gellissen, K., and Kowalewski, S.: Isolierter Defekt der tubulären Cystin-Rückresorption in einer Familie mit idiopathischen Hypoparathyroidismus, Klin. Wchnschr. 45:38-40, 1967.

76. Joseph, R., Ribierre, M., Job, J. C., and Girault, M.: Maladie familiale associant des convulsions à début très précoce une hyperalbuminorachie et une hyperaminoacidurie, Arch. franc. pediat. 15:374-387, 1958.

77. Sheldon, W., Luder, J., and Webb, B.: A familial tubular absorption defect of glucose and amino acids, Arch. Dis. Childhood 36:90-95, 1961.

78. Hooft, C., Timmermans, J., Snoeck, J., Antener, I., Oyaert, W., and Hende, Ch. V. D.: Methionine malabsorption syndrome, Ann. paediat. 205:73-104, 1965.

79. Kaser, H., Cottier, P., and Antener, I.: Glucoglycinuria, a new familial syndrome, J. Pediat. 61:386-394, 1962.

80. Scriver, C. R., Goldbloom, R. B., and Roy, C. C.: Hypophosphatemic rickets with renal hyperglycinuria, renal glucosuria and glycyl-prolinuria. A syndrome with evidence for renal tubular secretion of phosphorus, Pediatrics 34:357-371, 1964.

81. De Vries, A. D., Kochwa, S., Lazebnik, J., Frank, M., and Djaldetti, M.: Glycinuria, a hereditary disorder associated with nephrolithiasis, Am. J. Med. 23:408-415, 1957.

82. Scriver, C. R.: Familial Iminoglycinuria in Stanbury, J. B., Wyngaarden, J. B., and Fredrickson, D. S. (eds.): *The Metabolic Basis of Inherited Disease* (3rd ed.; New York: McGraw-Hill Book Company, Inc., 1972).

83. Kekomaki, M., Visakorpi, J. K., Perheentupa, J., and Saxen, L.: Familial protein intolerance with deficient transport of basic amino acids. An analysis of 10 patients, Acta paediat. Scandinav. 56:617-630, 1967.

84. Rowley, P. T., Mueller, P. S., Watkin, D. M., Rosenberg, L. E.: Familial growth retardation, renal amino aciduria, and cor pulmonale. I. Description of a new syndrome with case reports, Am. J. Med. 31:187-204, 1961.

85. Rosenberg, L. E., Mueller, P. S., and Watkin, D. M.: A new syndrome. Familial growth retardation, renal aminoaciduria, and cor pulmonale. II. Investigation of renal function, amino acid metabolism, and genetic transmission, Am. J. Med. 31:205-215, 1961.

86. Worthen, H. G., and Good, R. A.: The de Toni-Fanconi syndrome with cystinosis. Clinical and metabolic study of two cases in a family and a critical

review of the nature of the syndrome, Am. J. Dis. Child. 95:653-688, 1958.

87. Dent, C. E.: The amino-aciduria in Fanconi syndrome. A study making extensive use of techniques based on paper partition chromatography, Biochem. J. 41:240-253, 1947.

88. Cusworth, D. C., Dent, C. E., and Flynn, F. V.: Amino-aciduria in galactosaemia, Arch. Dis. Childhood 30:150-154, 1955.

89. Lampert, F., Mayer, H., Tocci, P. M., and Nyhan, W. L.: Fanconi syndrome in glycogen storage disease in Nyhan, W. L. (ed.): *Amino Acid Metabolism and Genetic Variation* (New York: McGraw-Hill Book Company, 1967).

90. Darling, S., Mortensen, O., and Sondergaard, G.: Lactosuria and amino-aciduria in infancy. A new inborn error of metabolism?, Acta paediat. 49:281-290, 1960.

91. Richards, W., Donnell, G. N., Wilson, W. A., and Stowens, D. S.: The oculo-cerebro-renal syndrome of Lowe, Am. J. Dis. Child. 100:707-709, 1960.

92. Abbassi, V., Lowe, C. U., and Calcagno, P. L.: Oculo-cerebro-renal syndrome. A review, Am. J. Dis. Child. 115:145-168, 1968.

93. Stein, W. H., Bearn, A. G., and Moore, S.: The amino acid content of the blood and urine in Wilson's disease, J. Clin. Invest. 33:410-419, 1954.

94. Bearn, A. G.: Wilson's disease. An inborn error of metabolism with multiple manifestations, Am. J. Med. 22:747-757, 1957.

95. Woolf, L. I., and Norman, A. P.: The urinary excretion of amino acids and sugars in early infancy, J. Pediat. 50:271-295, 1957.

96. Ghadimi, H., and Shwachman, H.: Evaluation of aminoaciduria in infancy and childhood, Am. J. Dis. Child. 99:457-475, 1960.

97. Ganis, F. M., and Howland, J. W.: Observations on amino-acid excretion in accidentally-irradiated humans, Internat. J. Radiat. Biol. 7:107-108, 1963.

98. Clarkson, T. W., and Kench, J. E.: Urinary excretion of amino acids by men absorbing heavy metals, Biochem. J. 62:361-372, 1956.

99. Gerritsen, T., Vaughn, J. G., and Waisman, H. A.: The origin of homocitrulline in the urine of infants, Arch. Biochem. Biophys. 100:298-301, 1963.

100. Dull, T. A., and Henneman, P. H.: Urinary hydroxyproline as an index of collagen turnover in bone, New England J. Med. 268:132-134, 1963.

101. Benoit, F. L., Theil, G. B., and Watten, R. H.: Hydroxyproline excretion in endocrine disease, Metabolism 12:1072-1082, 1963.

102. Feigin, R. D., Klainer, A. S., Beisel, W. R., and Hornick, R. B.: Whole-blood amino acids in experimentally induced typhoid fever in man, New England J. Med. 278:293-298, 1968.

103. Prasad, L. S., Rahman, A., Sen, D. K., et al.: Some observations on serum proteins and aminoaciduria in malnutrition, Indian Pediat. 6:73-83, 1969.

104. Gjessing, L. R.: Studies of functional neural tumors. II. Cystathioninuria, Scandinav. J. Clin. & Lab. Invest. 15:474-478, 1963.

105. Opienska-Blauth, J., and Kowalska, H.: Aminoaciduria in the nephrotic syndrome in children, Clin. chim. acta 6:805-813, 1961.

106. Frimpter, G. W., Timpanelli, A. E., Eisenmenger, W. J., Stein, H. S., and Ehrlich, L. I.: Reversible "Fanconi syndrome" caused by degraded tetracycline, J.A.M.A. 184:111-113, 1963.

107. Andrews, B. F., Bruton, O. C., and Knollock, E. C.: Amino-aciduria in salicylate intoxication, Am. J. Med. Sc. 242:411-414, 1961.

108. Eades, C. H., Jr., Pollack, R. L., and Hardy, J. D.: Thermal burns in man. IX. Urinary amino acid patterns, J. Clin. Invest. 34:1756-1759, 1955.

109. Jonxis, J. H. P., and Huisman, T. H. J.: Amino aciduria and ascorbic acid deficiency, Pediatrics 14:238-244, 1954.

110. Jonxis, J. H. P., Smith, P. A., and Huisman, T. H. J.: Rickets and amino-aciduria, Lancet 2:1015-1017, 1952.

111. Stevenson, R. E., and Huntley, C. C.: Congenital malformations in offspring of phenylketonuric mothers, Pediatrics 40:33-45, 1967.

112. Howell, R. R., and Stevenson, R. E.: The offspring of phenylketonuric women, Social Biology 18: (suppl.) 19-29, 1971.

113. Dent, C. E.: Applications to studies of amino acid and protein metabolism, Biochem. J. 43:XLVIII, 1948.

114. Ghadimi, H., and Shwachman, H.: A screening test for aminoaciduria, New England J. Med. 261:998-1001, 1959.

115. Efron, M. L., Young, D., Moser, H. W., and MacCready, R. A.: A simple chromatographic screening test for the detection of disorders of amino acid metabolism. A technic using whole blood or urine collected on filter paper, New England J. Med. 270: 1378-1383, 1964.

116. Scriver, C. R., Davies, E., and Cullen, A. M.: Application of a simple micromethod to the screening of plasma for a variety of aminoacidopathies, Lancet 2:230-232, 1964.

117. Berry, H. K., Leonard, C., Peters, H., Granger, M., and Chunekamrai, N.: Detection of metabolic disorders. Chromatographic procedures and interpretation of results, Clin. Chem. 14:1033-1065, 1968.

118. Moore, S., and Stein, W. H.: Chromatography of amino acids on sulfonated polystyrene resins, J. Biol. Chem. 192:663-681, 1951.

119. Spackman, D. H., Stein, W. H., and Moore, S.: Automatic recording apparatus for use in the chromatography of amino acids, Anal. Chem. 30:1190-1206, 1958.

120. Hamilton, P. B.: Ion exchange chromatography of amino acids, Anal. Chem. 30:914-919, 1958.

121. Hamilton, P. B.: Ion exchange chromatography of amino acids, Anal. Chem. 35:2055-2064, 1963.

122. Hamilton, P. B.: The significance of sensitivity and resolution in the ion-exchange chromatographic determination of amino acids in biologic fluids, Clin. Chem. 14:535-537, 1968.

123. Perry, T. L., Stedman, D., and Hansen, S.: A versatile lithium buffer elution system for single column automatic amino acid chromatography, J. Chromatog. 38: 460-466, 1968.

124. Samyn, W., Carton, D., and Hooft, C.: Column chromatography of amino acids in the study of metabolic diseases, Clin. chim. acta 28:83-88, 1970.

125. Efron, M.: Two way separation of amino acids and other ninhydrin-reacting substances by high-voltage electrophoresis followed by paper chromatography, Biochem. J. 72:691-694, 1959.

126. Mabry, C. C., and Todd, W. R.: Quantitative measurement of individual and total free amino acids in urine. Rapid method employing high-voltage paper electrophoresis and direct densitometry and its application to the urinary excretions of amino acids in normal subjects, J. Lab. & Clin. Med. 61:146-157, 1963.

127. Mabry, C. C., and Karam, E. A.: Measurement of free amino acids in plasma and serum by means of high-voltage paper electrophoresis, Am. J. Clin. Path. 42:421-430, 1964.

128. Sackett, D. L.: Adaptation of monodirectional high-voltage electrophoresis on long papers to the rapid qualitative identification of urinary amino acids, J. Lab. & Clin. Med. 63:306-314, 1964.

129. Zumwalt, R. W., Roach, D., and Gehrke, C. W.: Gas-liquid chromatography of amino acids in biological substances, J. Chromatog. 53:171-194, 1970.

130. Perry, T. L., and Hansen, S.: Technical pitfalls leading to errors in the quantitation of plasma amino acids, Clin. chim. acta 25:53-58, 1969.

131. Hamilton, P. B.: Amino acids on hands, Nature 205: 284-285, 1965.

132. Jepson, J. B.: Hartnup Disease in Stanbury, J. B., Wyngaarden, J. B., and Fredrickson, D. S. (eds.): *The Metabolic Basis of Inherited Disease* (3rd ed.; New York: McGraw-Hill Book Company, Inc., 1972).

Index